GEOGRAPHY
FOR EDEXCEL

A LEVEL YEAR 2

REVISION GUIDE

Series editor
Bob Digby

Catherine Hurst
Rebecca Priest
Andy Slater
Kate Stockings
Rebecca Tudor

OXFORD
UNIVERSITY PRESS

OXFORD
UNIVERSITY PRESS

Great Clarendon Street, Oxford, OX2 6DP, United Kingdom

Oxford University Press is a department of the University of Oxford.
It furthers the University's objective of excellence in research, scholarship,
and education by publishing worldwide. Oxford is a registered trade mark of
Oxford University Press in the UK and in certain other countries

© Oxford University Press 2018

Series editor: Bob Digby

Authors: Catherine Hurst, Rebecca Priest, Andy Slater, Kate Stockings,
Rebecca Tudor

The moral rights of the authors have been asserted.

Database right of Oxford University Press (maker) 2018.

First published in 2018

British Library Cataloguing in Publication Data
Data available

ISBN 978-019-843275-3

10 9 8 7 6 5 4

Paper used in the production of this book is a natural, recyclable product made
from wood grown in sustainable forests. The manufacturing process conforms
to the environmental regulations of the country of origin.

Printed and bound by CPI Group (UK) Ltd, Croydon, CR0 4YY

Acknowledgements
The publisher and authors would like to thank the following for permission
to use photographs and other copyright material:

Cover: reptiles4all/Shutterstock; **p54:** Thomas Cockrem/Alamy Stock
Photo; **p85:** Petar Kujundzic/REUTERS; **p88:** Bob Digby; **p93:** George Esiri/
REUTERS; **p101:** DIMITAR DILKOFF/Getty Images; **p110:** Nattapoom V/
Shutterstock; **p120:** NASA; **p125:** Reproduced by kind permission of
PRIVATE EYE magazine. Map created by suemarcar © OpenStreetMap
contributors © CARTO © Crown copyright Ordnance Survey; **p135:** Bob
Digby; **p136:** Ashley Cooper/Alamy Stock Photo; **p137, 138:** Bob Digby.

Artwork by Kamae Design and Aptara Inc.

Every effort has been made to contact copyright holders of material
reproduced in this book. Any omissions will be rectified in subsequent
printings if notice is given to the publisher.

Contents

Introduction: Aiming for success	5

1 The water cycle and water insecurity	16
Chapter overview	16
1.1 A world of extremes	17
1.2 The global hydrological cycle	18
1.3 Local hydrological cycles – drainage basins	20
1.4 The water balance and river regimes	22
1.5 Deficits within the hydrological system	24
1.6 Surpluses within the hydrological system	26
1.7 Climate change	28
1.8 Water insecurity – the causes	30
1.9 Water insecurity – the consequences	32
1.10 Water insecurity – managing supplies	34

2 The carbon cycle and energy security	36
Chapter overview	36
2.1 2015 – The year it all changed?	37
2.2 The carbon cycle	38
2.3 Carbon sequestration	40
2.4 A balanced carbon cycle	42
2.5 Energy security	45
2.6 Fossil fuels – still the norm	47
2.7 Alternatives to fossil fuels	49
2.8 Threats to the carbon and water cycles	51
2.9 Degrading the water and carbon cycles	53
2.10 Responding to climate change	55

3 Superpowers	57
Chapter overview	57
3.1 Making an impact	58
3.2 What is a superpower?	59
3.3 Changing patterns of power	61
3.4 Emerging superpowers?	63
3.5 Global networking	65
3.6 Players in international decision-making	67
3.7 Superpowers and the environment	69
3.8 Contested places	71
3.9 Contesting global influence	73
3.10 Challenges for the future	75

4 Health, human rights and intervention	77
Chapter overview	77
4.1 Human development	78
4.2 Variations in human health and life expectancy	80
4.3 Development targets and policies	82
4.4 Human rights	84
4.5 Defining and protecting human rights	86
4.6 Variations in human rights within countries	88
4.7 Geopolitical interventions	90
4.8 The impacts of development aid	92
4.9 Military action and human rights	94
4.10 Measuring the success of geopolitical interventions	96
4.11 Development aid – a mixed record of success?	98
4.12 Military interventions: a mixed success story	100

Contents

5	Migration, identity and sovereignty	102
Chapter overview		102
5.1	A national game?	103
5.2	Globalisation and migration	104
5.3	The causes of migration	106
5.4	The consequences of international migration	108
5.5	Nation states and borders	110
5.6	Nationalism in the modern world	112
5.7	Globalisation and the growth of new types of states	114
5.8	The role and importance of the United Nations	116
5.9	The role of IGOs in trade and finance	118
5.10	The role of IGOs in managing global environmental problems	120
5.11	The concept of national identity	122
5.12	Challenges to national identity	124
5.13	Disunity within nations	126

6	Preparing for synoptic Paper 3	128
6.1	Preparing for synoptic Paper 3	128
6.2	Using the Resource Booklet	130
6.3	Understanding the synoptic themes	131
6.4	Resource Booklet: Australia – can its growth be sustained?	133
6.5	Paper 3 exam-style questions	139
6.6	Understanding how Paper 3 is assessed	140
6.7	Exam-style questions mark scheme	142

Glossary	147
Revision planner	152

Introduction: Aiming for success

If you want to be successful in your exams, then you need to revise all you've learned for your A level course! That can seem daunting – but it's why this book has been written. It contains key points that you need to learn to prepare for exams for the Edexcel A Level Geography specification.

This Revision Guide is one of five publications from Oxford University Press to support your learning. The others are:

- *Geography for Edexcel A Level Year 2* student book, which this Revision Guide works alongside. All page links in this book refer to the student book.
- *Geography for Edexcel A Level* Year 1 and AS student book, for which there is also a separate revision guide.
- *A Level Geography for Edexcel Exam Practice.*

How to use this book

This Revision Guide contains the following to help you revise for the Edexcel A Level Geography specification.

An introduction to each of Papers 1–3

These contain outlines of:

- the three exam papers you'll be taking (pages 6–7)
- topics from the specification in each Paper.

Guidance on Paper 3

Exam Paper 3 differs from Papers 1 and 2. It consists of an unseen Resource Booklet, which assesses your ability to interpret and make sense of resources focused around five compulsory topics in the specification. Chapter 6 of this Revision Guide contains detailed advice to help you with Paper 3.

Content summaries

Each section in this Revision Guide summarises exactly the corresponding section in the student book, thus covering all the topics in the specification. Key content in the student book is summarised:

- each two-page topic in the student book is summarised on a single page in this Revision Guide
- each four-page topic is summarised on two pages.

Each section contains the following features:

You need to know – at the start of every section, this feature summarises key things you need to learn for each topic.

Main content – a summary of the main content found in the student book.

Ten-second summary – this summarises the essentials that you need to know, like a checklist.

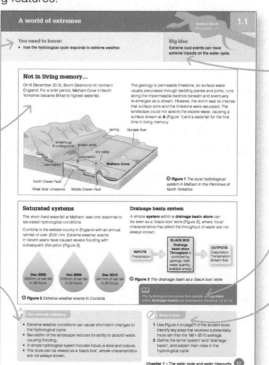

Big idea this spells out the big geographical ideas around which this A level has been written.

Cross-references – to the student book allow you to re-read topics in depth.

Over to you – these are activities focused on helping you learn material for the exam.

◀ *Figure 1* Your Revision Guide's key features

The Edexcel A Level Geography specification has eight topics, which together are assessed by three exams (Papers 1–3). Broadly, topics on physical geography are on Paper 1 and topics on human geography are on Paper 2. Paper 3 is a synoptic exam, designed to draw parts of the course together.

Paper 1

This paper has three sections, each assessing particular topics.

- **Section A** Tectonic Processes and Hazards
- **Section B Either** Glaciated Landscapes and Change **or** Coastal Landscapes and Change
- **Section C** The Water Cycle and Water Insecurity, **and** The Carbon Cycle and Energy Security

Marks and question types for each section are shown in Figure 2.

Section	Marks	Topics	Question	Details
A	16	Tectonic Processes and Hazards	Q1	• 1 x 4-mark question assessing quantitative skills • 1 x 12-mark essay question using the command word 'Assess'
B	40	**Either** Glaciated Landscapes and Change **or** Coastal Landscapes and Change	**Either** Q2 **or** Q3	• 2 x 6-mark paragraph questions using the command word 'Explain' • 1 x 8-mark longer paragraph question using the command word 'Explain' • 1 x 20-mark essay question using the command word 'Evaluate'
C	49	The Water Cycle and Water Insecurity **and** The Carbon Cycle and Energy Security	Q4	• 1 x 3-mark short paragraph question using the command word 'Explain' • 1 x 6-mark paragraph question using the command word 'Explain' • 1 x 8-mark longer paragraph question using the command word 'Explain' • 1 x 12-mark essay question using the command word 'Assess' • 1 x 20-mark essay question using the command word 'Evaluate'

⬆ **Figure 2** *Paper 1 topics, question styles and command words*

Paper 2

This paper also has three sections, each assessing particular topics.

- **Section A** Globalisation **and** Superpowers
- **Section B Either** Regenerating Places **or** Diverse Places
- **Section C Either** Health, Human Rights and Intervention **or** Migration, Identity and Sovereignty

Marks and question types for each section are shown in Figure 3.

Section	Marks	Topics	Question	Details
A	32	Globalisation **and** Superpowers	Q1 Q2	For Globalisation • **Either** 1 x 4-mark short paragraph question **or** 1 x 4 marks assessing quantitative skills • 1 x 12-mark essay question using the command word 'Assess' For Superpowers • **Either** 1 x 4-mark short paragraph question **or** 1 x 4 marks assessing quantitative skills • 1 x 12-mark essay question using the command word 'Assess'
B	35	**Either** Regenerating Places **or** Diverse Places	**Either** Q3 **or** Q4	• 1 x 3-mark short paragraph question using the command word 'Suggest' • 2 x 6-mark paragraph questions using the command words 'Suggest' and 'Explain' • 1 x 20-mark essay question using the command word 'Evaluate'
C	38	**Either** Health, Human Rights and Intervention **or** Migration, Identity and Sovereignty	**Either** Q5 **or** Q6	• **Either** 1 x 4 mark short paragraph question **or** 4 marks for quantitative skills • 1 x 6-mark paragraph question using the command word 'Explain' • 1 x 8-mark longer paragraph question using the command word 'Explain' • 1 x 20-mark essay question using the command word 'Evaluate'

△ **Figure 3** *Paper 2 topics, question styles and command words*

Paper 3

Paper 3 is a synoptic paper in the form of an issue analysis, presented as an unseen Resource Booklet on which questions are set. The topic will be drawn from the five compulsory topics in the specification, i.e.

• Tectonic Hazards (Topic 1)
• Globalisation (Topic 3)
• The Water Cycle and Water Insecurity (Topic 5)
• The Carbon Cycle and Energy Security (Topic 6)
• Superpowers (Topic 7).

Marks and question types for each section are shown in Figure 4.

Section	Marks	Question	Details
A	12	Q1–3	• 1 x 4-mark short paragraph question using the command word 'Explain' • 1 x 4-mark question assessing quantitative skills • 1 x 4-mark short paragraph question using the command words 'Explain' or 'Suggest'.
B	16	Q4–5	• 2 x 8-mark longer paragraph questions using the command word 'Analyse' (based on data in the Resource Booklet)
C	42	Q6–7	• 1 x 18-mark essay question using the command word 'Evaluate' (based largely on the Resource Booklet) • 1 x 24-mark essay question using the command word 'Evaluate' (based on the Resource Booklet in a wider context)

△ **Figure 4** *Paper 3 question styles and command words*

Command words and marks

In order to assess students of different abilities, examiners use an 'incline of difficulty' in each exam paper. This means the questions early in each section count for fewer marks and are more straightforward than those that appear later with higher marks.

To aid this, examiners use the command words shown in Figure 5. Each carries a certain number of marks, as shown in Figure 6. Some command words are designed to be more challenging than others. Command words are used consistently throughout the three exam papers, so a question using the command word 'Assess' carries 12 marks, whether it is in Paper 1 or Paper 2.

	Command word	Definition
Low-tariff questions	**Calculate**	Produce a numerical answer, using relevant working.
	Draw/Plot	Create a graphical representation of geographical information.
	Complete	Create a graphical representation of geographical information by adding detail to a resource that's provided.
	Suggest	For an unfamiliar scenario, provide a reasoned explanation of how or why something may occur.
	Explain	Provide a reasoned explanation of how or why something occurs. An explanation requires a justification/exemplification of a point.
Medium-tariff questions	**Analyse** (only used in Paper 3)	Break something down into individual components/processes and say how each contributes to the question's theme/topic and how components/processes work together.
High-tariff questions	**Assess**	Use evidence to determine the relative significance of something. Give balanced consideration to all factors and identify which are the most important.
	Evaluate	Measure the value or success of something and ultimately provide a balanced and substantiated judgement/conclusion. Review information and bring it together to form a conclusion, drawing on evidence, e.g. strengths, weaknesses, alternatives and relevant data.

🔺 **Figure 5** *Command words used in Edexcel A Level Geography exams*

Command word	3	4	6	8	12	18	20	24
Calculate		*						
Draw/Plot/Complete		*						
Suggest	*	*	*					
Explain	*	*	*	*				
Analyse (Paper 3 only)				*				
Assess					*			
Evaluate						* Paper 3 only	* Papers 1 and 2 only	* Paper 3 only

🔻 **Figure 6** *Marks used for each command word*

Although the command words are used consistently across the exam papers, there are some slight variations.

- **Calculate** – normally requires a process of calculation, with marks awarded for the process of reaching an answer, as well as the answer itself.
- **Suggest** – is used where you are given an unseen resource in Papers 1 and 2, and asked to suggest what the reasons might be for something. You are being marked for your reasoning, on the basis of having studied Geography at A level.
- **Explain** – is used slightly differently in 6- and 8-mark questions.
  - With 6-mark questions, 'Explain' requires you to interpret a resource or use a resource to help stimulate your thinking.
  - With 8-mark questions, you are expected to know the answer without any stimulus material.
- **Evaluate** – is used for 20-mark questions in Papers 1 and 2, but for 18- and 24-mark questions in Paper 3. It's simply a device for asking questions differently. The meaning, and the degree of challenge, is still the same, and examiners would argue that it is the most challenging command word in any question.

Why should you have to understand assessment objectives?

An assessment objective (AO) is the key tool used by examiners to decide what you should know, should understand and be able to do after studying Geography A Level for two years. Read this section to get a clear idea of why it will help you to know and understand how you're being assessed, as well as the topics you have to learn.

What are the assessment objectives?

There are three assessment objectives in Geography A Level:

- **AO1** is about your **knowledge and understanding**. It might be your knowledge and understanding of places, processes or issues. It's basically the content of the two student books in this series!
- **AO2** is about the way in which you **interpret and apply** your understanding to situations. For example, you could be asked to consider an argument. Imagine a question such as *'Can child labour ever be justified?'* You'd have to **weigh up evidence** (your knowledge and understanding: AO1) and frame it together into an **argument**. You might find there are points you could make that could support the use of child labour in certain circumstances, followed by others against its use. By the end, you might be able to **make a judgment** – perhaps in favour, perhaps against. That process – of using information to develop an argument – is what AO2 is all about. Most high-mark exam questions that you'll answer will have large numbers of marks allocated to AO2 – it's an essential A level skill.

- **AO3** is about using **geographical skills** in formulating questions, in thinking about methods of data collection, in manipulating and presenting data, and in drawing conclusions. You should recognise that this is exactly what you've been doing in writing up your individual investigation, known as the Non-Examined Assessment (NEA). You will have to use those same skills in some situations in the exam as well.

- **AO1** questions use the command word 'Explain'. You also need knowledge and understanding for exam questions assessing AO2.
- **AO2** questions use the command words 'Suggest', 'Assess' or 'Evaluate', requiring you to use your knowledge and understanding (AO1).
- **AO3** questions use the command words 'Draw', 'Plot', 'Complete' or 'Calculate' (for statistics questions) or 'Analyse' (for questions about data interpretation).

	Objective	%	Marks
AO1	Demonstrate knowledge and understanding of places, environments, concepts, processes, interactions and change, at a variety of scales	34%	119
AO2	Apply knowledge and understanding in different contexts to interpret, analyse and evaluate geographical information and issues	40%	139
AO3	Use a variety of relevant quantitative, qualitative and fieldwork skills to: • investigate geographical questions and issues • interpret, analyse and evaluate data and evidence • construct arguments and draw conclusions	26%	92

▲ *Figure 7* Assessment objectives used in the exams

Marks allocated to assessment objectives

It helps to prepare for each exam if you know which AOs are being assessed (Figure 8 on page 10).

- For example, Papers 1 and 2 almost entirely assess **AO1** (knowledge and understanding) and **AO2** (application). This means that you really need to know

your material, because AO1 counts for 46 of the total 105 marks, and AO2 for 55.
- Similarly, Figure 8 shows that Paper 3 (with the unseen Resource Booklet) has more **AO3** marks. That's because some questions will assess your ability to read, manipulate and interpret data in the booklet.

However, there are also a substantial number of **AO2** marks, because you'll be asked to analyse and make judgments about some of the issues, and **AO1** marks, because you'll be asked to relate what the issue is about to other topics you've learned.

	AO1 marks	AO2 marks	AO3 marks	Total
Paper 1	46	55	4	**105**
Paper 2	46	55	4	**105**
Paper 3	19	21	30	**70**
NEA	8	8	54	**70**
Total	**119**	**139**	**92**	**350**

⬥ **Figure 8** *Marks used for each assessment objective in the exams*

Command words, assessment objectives and marks

You know that some command words are more challenging than others. Figure 9 shows this in more detail – 'Explain' may carry up to 8 marks, while 'Evaluate' may carry 20.

However, several command words, shown in Figure 9, assess more than one assessment objective.

- For example, in Papers 1 and 2, you may be given a resource (e.g. Figure 10) and be asked to explain something about it for 6 marks.
- Imagine the question, '*With reference to Fig. 10, explain why the three river regimes might vary.*'
- To do this, you need to recognise patterns in the river regime and know something about why they vary, so 3 of the 6 marks are for AO1. The remaining marks are for applying what you know. You might know several possible reasons (e.g. snow melt, monsoon rainfall) but only some of these might apply. You therefore need to select the appropriate information – that's AO2.

As marks increase, so AO2 becomes more important.

- Figure 9 shows that 20-mark questions using 'Evaluate' are split – 5 marks for AO1 and 15 marks for AO2.
- A question such as '*Evaluate the factors responsible for a country's energy security*' would therefore carry 5 marks for knowledge and understanding and 15 marks for AO2. (On page 14, you can see how examiners mark responses like this.)

Command word	Total mark	AO1	AO2	AO3
Explain	6	6		
Explain (using a resource)	6	3	3	
Explain	8	8		
Assess	12	3	9	
Papers 1 and 2 Evaluate	20	5	15	
Paper 3 – Analyse	8	4		4
Paper 3 – Evaluate	18	3	9	6
Paper 3 – Evaluate	24	4	12	8

⬥ **Figure 9** *The balance of assessment objectives in questions using particular command words*

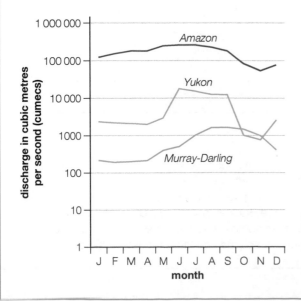

◀ **Figure 10** *Discharge patterns in three complex river regimes measured at the mouth of each river. This could be used as a resource for examiners to test your knowledge (AO1) and whether you can apply that knowledge to the graph (AO2)*

You probably remember learning several case studies for GCSE and it's possible you learned a lot of detail off by heart. A level is different. You need examples but not huge volumes of detail. It's much more important that you learn to argue a case based on concepts, with enough detail to support your case.

In the student book, Superpowers is one of the most popular A level topics. Here are three stages to help you learn the essentials:

Examples you need to know for Superpowers

- An example of contested political spheres of influence and tensions over territory and physical resources (e.g. South and East China Seas)
- An example of the potential for open conflict over territory and physical resources (e.g. Western Russia / Eastern Europe)
- An example of the rising economic importance of certain Asian countries (e.g. China and India)

1 Build up a factfile

A basic **factfile** about a conflict could include:

- Location – Where is it? Can you locate it?
- What kind of conflict is it?
- When did it occur? Date and time?
- Was it a single event or one of several?
- Describe *briefly* what happened, e.g. What was the conflict between Russia and Ukraine about? Who did what?
- What were the consequences?

2 Know its impacts

The impacts of conflict can be significant. What were the short-, medium- and long-term impacts? Be clear that you know what this means.

- **Short-term** – within the first month.
- **Medium-term** – within six months.
- **Long-term** – anything longer than six months.

Next, classify these into economic, social or environmental impacts.

- **Economic impacts** – e.g. jobs, businesses, trade, costs.
- **Social impacts** – people, health and housing.
- **Environmental impacts** – changes to the surrounding landscape.

You can now list these impacts and classify them, using a table like Figure 11. You need two impacts in each table 'cell'.

Impact	Immediate / short term	Medium term	Long term
Economic			
Social			
Environmental			

Figure 11 *A table for classifying the impacts of a conflict between countries*

3 Know the key ideas

Key ideas are particularly important because it's these that carry most marks. Superpowers is assessed by a 12-mark question using the command word 'Assess'. Study the examples of key ideas in the panel below and try to design your own exam questions. An example for Key idea 4 could be: '*Assess the extent to which global influence can be contested in a range of different economic, environmental and political spheres.*'

For each one, you could then draw a mind map to show how you might answer the question.

Key ideas you need to understand for Superpowers

1 The human and physical of characteristics of superpowers
2 Patterns of power change over time
3 Emerging powers vary in their influence on people and the physical environment
4 How global influence is contested in a number of different economic, environmental and political spheres
5 Superpowers play a key role in international decision-making
6 Developing nations have changing relationships with superpowers.

You'll spend most of your time for preparing for the exam revising content for each topic. You'll also need to prepare for Paper 3 by revising details about players, attitudes and actions, and futures and uncertainties (see Chapter 6 in this Revision Guide).

However, you should also be aware of, and do some preparation to understand, the 14 key concepts (Figure 12) on which the whole A level is based. All A levels in England and Wales are based on these. Examiners could use the key concepts in any question and expect you to be able to say something about them.

Over to you

Copy and complete the table with examples in the right-hand column.

Concept	Definition	Example
Causality	Connections between cause and consequence as part of a process	
Equilibrium	A condition in which all influences acting within a system cancel each other or self-correct, so that systems are balanced	
Feedback	Responses to change; positive feedback causes further change and instability to a system, while negative feedback returns a system to equilibrium.	
Globalisation	A process leading to greater international integration economically, culturally and demographically	
Identity	The beliefs, perceptions and characteristics that make one group of people or places seem different to others	
Inequality	Differences in opportunity, access to resources or outcomes (e.g. health) between different groups, at any scale	
Interdependence	Mutual reliance between groups; strongly linked to globalisation	
Mitigation and adaptation	Alternative approaches to management: preventative (mitigation) versus coping with (adaptation)	
Representation	How places or situations are portrayed to others, e.g. through stories, news items, photos, painting or other media	
Resilience	The ability to cope with change, e.g. resilience to global warming or to a hazard	
Risk	The potential or probability of harm / losing something of value	
Sustainability	The Brundtland definition refers to sustainability as *'development that meets the needs of the present without compromising the ability of future generations to meet their own needs'*, a contested term that can be interpreted in terms of economics, society, environment or politics.	
Systems	Many interacting component parts, producing a complex 'whole', with inputs, flows, stores and outputs	
Threshold	A critical level in a system beyond which change is inevitable/irreversible	

Figure 12 *The 14 key concepts*

	The water cycle and water insecurity	The carbon cycle and energy security	Superpowers	Health, human rights and intervention	Migration, identity and sovereignty
Causality					
Equilibrium					
Feedback					
Globalisation					
Identity					
Inequality					
Interdependence					
Mitigation and adaptation					
Representation					
Resilience					
Risk					
Systems					
Sustainability					
Threshold					

Figure 13 *Which key concepts apply to the chapters in this Revision Guide*

Over to you

Copy and complete Figure 13 with ticks (√) to show which of the 14 key concepts apply to the five chapters in this Revision Guide.

However you look at it, revision can be dull! But whatever revision you do should be **active**.
This page will help you develop useful ways of revising.

Revising in groups

Working as a group is always better than alone. Try these ideas out.

Form a study group with friends Fix time with two or three friends to go through key topics. Do timed questions together, then mark them. Make lists of things you don't understand to ask your teacher.

Working together at home Message, Facetime or Skype friends and test each other. Go through questions together.

Test each other Make flash cards of key words and construct mind maps of key concepts.

Go through the specification and get to know its key words and concepts (Figure 13).

Revising, using the specification

Figure 14 shows an extract from the specification. Use it to make sure you revise everything that you should. The important features are:

Key ideas Examiners will use these to construct the longer 12- or 20-mark questions.

Detailed content Examiners will use this for the shorter questions.

Globe symbol Examiners can expect you to be able to use examples in questions set on parts of the specification with a globe symbol.

Synoptic links These will be especially important in preparing for Paper 3.

This column contains the key ideas that will be assessed in longer exam questions, e.g. 'Assess the extent to which...'.

Broad enquiry questions form the basis of each sub-section of the specification.

This column contains all the key words and phrases you need to know. Most of the shorter 3-, 6- or 8-mark questions will come from this column.

The globe symbol means you should be taught examples, though any examples will do, not necessarily those shown.

The parts in bold are the synoptic links – P for players, A for attitudes and actions, F for futures and uncertainties.

Enquiry question 3: How does water insecurity occur and why is it becoming such a global issue for the 21st century?	
Key idea	Detailed content
5.7 There are physical causes and human causes of water insecurity.	a. The growing mismatch between water supply and demand has lead to a global pattern of water stress (below 1700 m³ per person and water scarcity (below 1000 m³ per person. (7)
	b. The causes of water insecurity are physical (🌐 climate variability, salt water encroachment at coast) as well as human (🌐) over abstraction from rivers, lakes and groundwater aquifers, water contamination from agriculture, industrial water pollution).
	c. The finite water resource faces pressure from rising demand (increasing population, improving living standards, industrialisation and agriculture), which is increasingly serious in some locations and is leading to increasing risk of water insecurity. **(F: projections of future water scarcity)**
5.8 There are consequences and risks associated with water insecurity.	a. The causes of global pattern of physical water scarcity and economic scarcity and why the price of water varies globally. (8)
	b. The importance of water supply for economic development (industry, energy supply, agriculture) and human wellbeing (sanitation, health and food preparation); the environmental and economic problems resulting from inadequate water.

▶ **Figure 14** How different parts of the specification are used by examiners

To gain the best marks possible in your exams, follow this advice.

1 Break down the question

Look at the example below (Figure 15). Try to break up questions like this. It will help you to focus on what the examiner is asking.

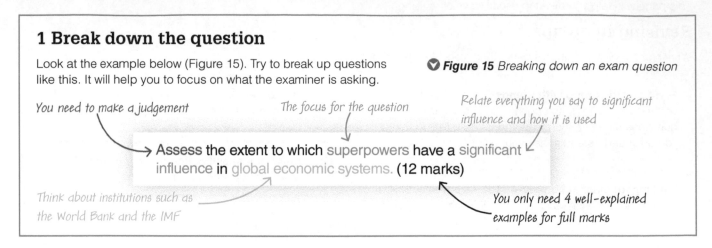

Figure 15 *Breaking down an exam question*

You need to make a judgement

The focus for the question

Relate everything you say to significant influence and how it is used

→ Assess **the extent to which** superpowers **have a** significant influence **in** global economic systems. **(12 marks)**

Think about institutions such as the World Bank and the IMF

You only need 4 well-explained examples for full marks

2 Plan your answer

Planning your answer helps to organise your thoughts. Some people plan using a spider diagram, others just make a short list. Planning helps you to get the order of the answer right and makes sure you don't forget what to write. Figure 16 is an example of a plan.

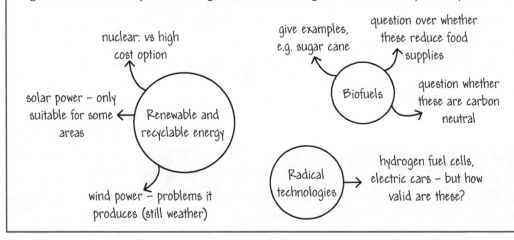

nuclear: vs high cost option

give examples, e.g. sugar cane

question over whether these reduce food supplies

solar power – only suitable for some areas

Renewable and recyclable energy

Biofuels

question whether these are carbon neutral

wind power – problems it produces (still weather)

Radical technologies

hydrogen fuel cells, electric cars – but how valid are these?

Figure 16 *An example of a plan for the 12-mark question, 'Evaluate the extent to which there are realistic alternatives to fossil fuels.' You could add more details to this plan*

3 Get to know the mark scheme

Questions of 6 marks or more are assessed using level descriptors. These are the qualities that examiners look for in an answer. They are a series of criteria against which your answer will be judged. Between 6 and 12 marks, there are three levels – Level 3 is the highest. Above 12 marks, there are four levels, with Level 4 the highest.

As an example, the Level 4 descriptors for 20-mark questions, which use the command word 'Evaluate', include the following:

- Demonstrates accurate and relevant geographical knowledge and understanding throughout – *i.e. you know your material well* (AO1)
- Applies knowledge and understanding of geographical information/ideas to find logical and relevant connections/relationships – *i.e. you sequence*

your ideas so that geographical patterns are clear in your argument, rather than 'whatever comes next' (AO2)

- Applies knowledge and understanding of geographical information/ideas to produce a full and coherent interpretation that is supported by evidence – *i.e. your examples are planned into a conceptual argument and you use evidence selectively to prove a point rather than write down 'all I know about …'* (AO2)
- Applies knowledge and understanding of geographical information/ideas to come to a rational, substantiated conclusion, fully supported by a balanced argument that is drawn together coherently – *i.e. you argue your points well and sustain your argument from start to finish* (AO2)

You'll often hear students say 'Good luck' to each other as they enter the exam room. If you have done certain things, you won't need luck. Exam success comes from following a few rules. Students who perform well almost always follow these rules.

They revise. Lack of revision always catches up with you. A levels demand knowledge and understanding of a great deal of content, so it's important to know your stuff!

They know which **topics** will be in each exam – for example, which exam tests physical or human geography, and which of their topics, such as tectonic processes and hazards, are on which paper.

They look at the **marks**, and know what sort of questions carry the highest marks. Similarly, they know and understand each **command word**.

They **practise** answers, often under timed conditions, for example, allowing 25 minutes for a 20-mark question.

They get **timing** right. Each paper is 2 hours and 15 minutes in length, but question length varies from one paper to another. For example, in Paper 3, you must take into account the time needed to read the Resource Booklet thoroughly. As a general rule on Papers 1 and 2, allow 13 minutes for every 10 marks.

They **answer everything that they should**, leaving no blanks. Even if unsure, they write something. Leaving a 12-mark answer blank could mean giving up a whole grade.

They write in **full sentences**. Single words or phrases are fine for 1–2 mark questions or for some of the quantitative skills questions. But 6-mark answers written in bullet points rarely score well and 20-mark 'Evaluate' answers even less so.

They learn **specific details** about case studies or examples. They take time to learn one or two statistics, names of places, and schemes. They don't just say 'in Africa'! Use specific place knowledge – you need this to earn the highest marks.

They get to know the **mark scheme** and the assessment objectives. They also get to know the **criteria** on which answers are judged and the differences between Levels 1, 2 and 3, or 4 where there is a Level 4 (see opposite).

Finally, they make sure they have a **timetable** that tells them exactly what time and which day each exam is on! Make sure you have one. Check and double-check it!

Chapter 1
The water cycle and water insecurity

What do you have to know?

This chapter studies the water cycle on a variety of spatial scales and timescales. It studies how physical processes control the circulation of water between land, oceans, the cryosphere and the atmosphere. It also focuses on water insecurity as a global issue with serious consequences and with a range of different approaches to its management.

The specification is framed around three enquiry questions:

1 What are the processes operating within the hydrological cycle from global to local scale?
2 What factors influence the hydrological system over short- and long-term timescales?
3 How does water insecurity occur and why is it becoming such a global issue for the 21st century?

The table below should help you.

- Get to know the key ideas. They are important because the 12- and 20-mark questions are likely to be based on these.
- Copy the table and complete the key words and phrases by looking at Topic 5 in the specification. Section 5.1 has been done for you.

Key idea	Key words and phrases you need to know
5.1 The global hydrological cycle is of enormous importance to life on Earth.	global hydrological cycle, a closed system (inputs, outputs, stores and flows), water stores (oceans, atmosphere, biosphere, cryosphere, groundwater, surface water), fluxes, global water budget, residence times, non-renewable stores (fossil water, cryosphere)
5.2 The drainage basin is an open system within the global hydrological cycle.	
5.3 The hydrological cycle influences water budgets and river systems at a local scale.	
5.4 Deficits within the hydrological cycle result from physical processes but can have significant impacts.	
5.5 Surpluses within the hydrological cycle can lead to flooding, with significant impacts for people.	
5.6 Climate change may have significant impacts on the hydrological cycle globally and locally.	
5.7 There are physical and human causes of water insecurity.	
5.8 There are consequences and risks associated with water insecurity.	
5.9 There are different approaches to managing water supply, some more sustainable than others.	

You need to know:
- how the hydrological cycle responds to extreme weather.

Big idea
Extreme local events can have extreme impacts on the water cycle.

Not in living memory...

On 6 December 2015, Storm Desmond hit northern England. For a brief period, Malham Cove in North Yorkshire became Britain's highest waterfall.

The geology is permeable limestone, so surface water usually percolates through bedding planes and joints, runs along the impermeable bedrock beneath and eventually re-emerges as a stream. However, the storm was so intense that surface soils and the limestone were saturated. The landscape could not absorb the excess water, causing a surface stream at **A** (Figure 1) and a waterfall for the first time in living memory.

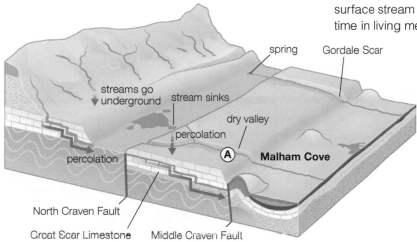

◄ **Figure 1** *The local hydrological system in Malham in the Pennines of North Yorkshire*

Saturated systems

The short-lived waterfall at Malham, was one response to saturated hydrological conditions.

Cumbria is the wettest county in England with an annual rainfall of over 2000 mm. Extreme weather events in recent years have caused severe flooding with subsequent disruption (Figure 2).

 Figure 2 *Extreme weather events in Cumbria*

Dec 2005 200 mm of rain fell in 36 hours

Nov 2009 316 mm of rain fell in 24 hours

Dec 2015 341 mm of rain fell in 24 hours

Drainage basin system

A simple **system** within a **drainage basin store** can be seen as a 'black box' store (Figure 3), where 'local' characteristics that affect the throughput of water are not always known.

INPUTS Precipitation

BLACK BOX Drainage basin store Throughput is controlled by geology, relief, water quantity, available energy

OUTPUTS Evaporation Transpiration Stream flow

 Figure 3 *The drainage basin as a 'black box' store*

The hydrological processes that operate as a **system** within **drainage basins** are examined in Sections 1.2 to 1.4.

Ten

Ten-second summary
- Extreme weather conditions can cause short-term changes to the hydrological cycle.
- Saturation of the landscape reduces its ability to absorb water, causing flooding.
- A simple hydrological system includes inputs, a store and outputs.
- The store can be viewed as a 'black box', whose characteristics are not always known.

Over to you
1 Use Figure 4 on page 7 of the student book. Identify key areas that received substantially more rain than the 1981–2010 average.
2 Define the terms 'system' and 'drainage basin', and explain their roles in the hydrological cycle.

You need to know:

- that the global hydrological cycle is a closed system
- the importance and size of stores and annual fluxes between atmosphere, ocean and land
- that the global water budget limits the availability of water
- about residence times and non-renewable stores.

Big idea

The global hydrological cycle is fundamental to life on Earth.

The global hydrological cycle

Freshwater makes up just 2.5% of all the water on Earth. **Residence (storage) times** of freshwater vary from one week to 10 000 years (Figure 1). **Solar energy** causes evaporation of water, which then returns as precipitation.

The global hydrological cycle is a **closed system**. In Figure 2, the total **inputs** (purple, large down arrows) are the same as the total **outputs** (blue, large up arrows).

- Inputs are dependent on outputs.
- Some **stores** are depleting, e.g. polar ice and glaciers.
- Evaporation is greatest in warm areas. Global air circulation takes the vapour to cooler areas where it condenses to form clouds and precipitation.

Two processes drive the global hydrological cycle:

Solar energy

- Energy from the sun heats water.
- Evaporation increases as the global climate warms.
- Moisture levels in the atmosphere increase.
- Condensation and precipitation increase as air cools.

Gravitational potential energy

- Water moves by gravity.
- Runoff and groundwater flow transport water to the sea (Figure 2).

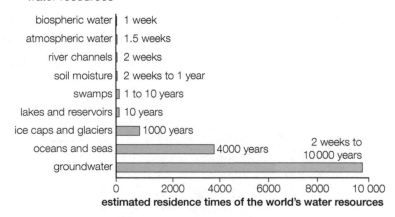

🔻 **Figure 1** *Estimated residence (storage) times of the world's water resources*

Water resource	Residence time
biospheric water	1 week
atmospheric water	1.5 weeks
river channels	2 weeks
soil moisture	2 weeks to 1 year
swamps	1 to 10 years
lakes and reservoirs	10 years
ice caps and glaciers	1000 years
oceans and seas	4000 years
groundwater	2 weeks to 10 000 years

estimated residence times of the world's water resources

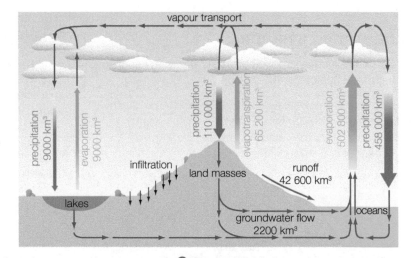

🔺 **Figure 2** *The global hydrological cycle*

Global stores and flows

Most freshwater is locked up in the **cryosphere** (ice – glaciers and ice sheets) or as groundwater. Only 0.4% of freshwater is in rivers, lakes, the atmosphere or the biosphere.

Global transfer of water is by flows known as **fluxes** (Figure 3). The variation due to seasons and temperature is the **annual flux**.

Store to store	Process	Flux amount of water (km³ per year)
Oceans and atmosphere	Evaporation	400 000
	Precipitation	370 000
Landmasses and atmosphere	Evaporation	60 000
	Precipitation	90 000
Landmasses and oceans	Surface runoff	30 000

🔺 **Figure 3** *Annual fluxes of water between global stores*

The global water budget

The **global water budget** is the difference (**balance**) between inputs and outputs.

- Oceans lose more water (outputs – evaporation) than they gain (inputs – precipitation).
- Land masses gain more water than they lose (Figure 3). Surface runoff makes up the balance (Figure 2).
- Residence time in the atmosphere is much shorter than in the oceans.

The importance of the Tropics

The steep angle of the sun at the Tropics intensifies radiation so there is high evaporation. Vapour is transferred towards the **Inter-Tropical Convergence Zone (ITCZ)**, where air rises and cools due to **convectional** currents, and forms clouds. The ITCZ is the biggest flux, transferring water from oceans to land.

The importance of the polar regions

The **cryosphere** locks up 66% of Earth's freshwater. As the climate warms, this is released into the sea.

Polar regions contribute to global circulation of water and the transfer of heat, which drive the **thermohaline circulation** (Figure 4).

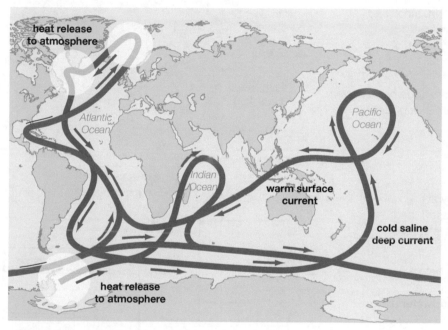

The hydrology of polar and tropical rainforest regions can be found on page 9 of the student book.

1 Polar ocean water is colder and denser than tropical water, so it sinks.
2 The sinking cold water draws in warm water from above, and subsequently draws water from the Tropics.
3 Water movement from the Tropics draws cold water from ocean depths to be warmed again.

Figure 4 *The thermohaline circulation – the flow of hot and cold water around the oceans*

Fossil water

Ancient stores of water beneath polar regions and deserts can be exploited due to new technology. **Aquifers** beneath deserts in Kenya contain 70 years' worth of freshwater at current usage.

Ten-second summary

- The global hydrological cycle is a closed system driven by solar energy and gravitational potential energy.
- The hydrology of different regions varies because of differences in inputs, transfers and water flows.
- The global water budget balances amounts of water received by oceans and continents.
- The tropics and polar regions are important in driving the global circulation of water.

Over to you

1 Sketch the global hydrological cycle and label its component parts.
2 Annotate the sketch to show:
   a how the cycle is a **closed system**
   b how and why inputs depend on outputs
   c why some stores are depleting.
3 Explain the links between the global hydrological cycle and the thermohaline circulation.

You need to know:
- that the hydrological cycle is a system of linked processes
- that physical factors determine the significance of inputs, flows and outputs
- that humans impact on the drainage basin cycle.

Big idea

The drainage basin is an open system within the hydrological cycle.

Open systems

An area drained by a river and its tributaries is a **drainage basin**, within which local hydrological processes operate (Figure 1). They are also known as **catchment areas** as they 'catch' the precipitation falling within a **watershed**.

Unlike the global hydrological system, drainage basins are **open systems**. This means that:

- Inputs are not determined by the outputs of the system.
- Open systems can lose more water than they receive – by evaporation and evapotranspiration, runoff and percolation.

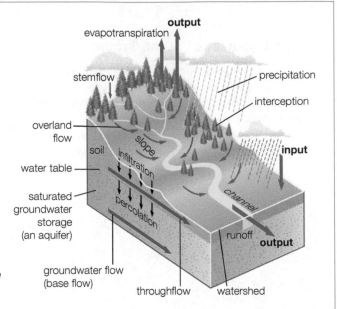

> **Figure 1** *The drainage basin system and the processes acting upon it*

Hydrological processes

Precipitation can follow just three pathways – **infiltration**, **overland flow** (**surface runoff**) and **evaporation** (Figure 2). These pathways can be delayed by:

- **interception** by plants or buildings before evaporation or infiltration into the surface
- **percolation** through rocks as **groundwater** and subsequent storage in **aquifers**.

Terminology is crucial in explaining the hydrological cycle. See page 13 of the student book for definitions of the key terms you need to know.

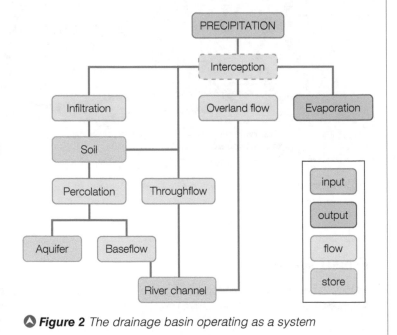

> **Figure 2** *The drainage basin operating as a system*

Drainage basin factors

Natural movement of water within a drainage basin is controlled by solar energy and gravity. Different factors control how precipitation reacts on reaching the land surface. These are **basin-wide factors** and include:

- relief
- geology
- climate
- land use
- vegetation
- drainage density.

The precipitation input

When warm moist air rises, it cools and condenses to form clouds, resulting in precipitation. The UK experiences three types of rain (Figure 3), in which air is forced to rise in three different ways:

- Drainage basins in western UK are exposed to warmer moist air masses from the Atlantic and are prone to **orographic** rainfall (also known as **relief** rainfall because it falls over high ground) or **frontal** rainfall (associated with low pressure). They tend to have high **water tables** and **antecedent** moisture, where water from one storm hasn't cleared by the time the next arrives.
- Because the western hills force orographic rain to fall in the west and north, eastern UK lies in a **rain shadow**, making it drier.
- In summer, in the drier drainage basins in eastern UK, the warm ground heats air above it, which rises, causing **convectional** air instability and thunderstorms. This can lead to **flash floods** and rapid runoff on dry soils.

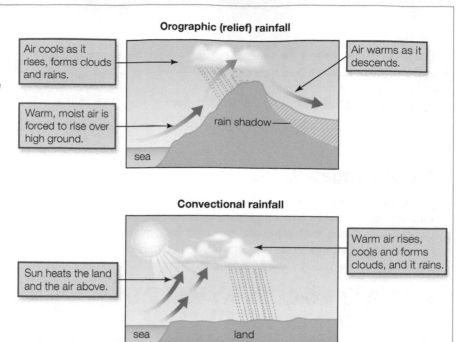

Orographic (relief) rainfall

Air cools as it rises, forms clouds and rains.

Warm, moist air is forced to rise over high ground.

Air warms as it descends.

rain shadow

sea

Convectional rainfall

Warm air rises, cools and forms clouds, and it rains.

Sun heats the land and the air above.

sea   land

Frontal rainfall

Air cools and forms clouds.

front

Heavy rain falls along the front.

Warm air is forced to rise over cold air.

> **Figure 3** Types of UK rainfall

Human impacts on drainage basins

Human activities have impacts on drainage basins (Figure 4).

Human activity	Impact
Over-abstraction	Abstracting too much water from groundwater can lead to rivers drying up during low rainfall.
Deforestation	Tropical forests flourish on thin soils, so removal of forest canopy has devastating effects on topsoils, natural ecosystems and the natural water cycle.
Change of land use, e.g. urbanisation	More impermeable surfaces mean increased runoff and reduced infiltration and evapotranspiration. Increased demand for water means more storage is needed and more abstraction from groundwater.
Reservoirs	Building new reservoirs or increasing abstraction meets demand but changes the natural water cycle by delaying the flow of water and removes water from the drainage basin. More water evaporates globally from reservoirs than is used by people.

Figure 4 Human impacts on drainage basins

You need to know:

- that water budgets show the annual balance between inputs and outputs
- that the ability of soil to hold moisture is vital
- what influences a river's regime
- the physical and human factors that affect a storm hydrograph's shape.

Big idea

The hydrological cycle influences water budgets and river systems at a local scale.

Water budgets

Water availability varies day-to-day and, over a longer period, the differences between inputs and outputs in any given area balance each other out. This balance is called the **water budget**. Inputs and outputs within the drainage basin control how much water is available (Figure 1).

Inputs	Outputs
• Precipitation • Water diversion into the area • Groundwater flow into the area • Surface water flow into the area • Surface runoff into the area	• Evapotranspiration • Water diversion out of the area • Groundwater flow out of the area • Surface water flow out of the area • Surface runoff out of the area • Industrial or residential uses within the area

🔺 **Figure 1** *Key components of the water budget that affect hydrological processes*

The water balance equation

The balance of total precipitation (**P**) in terms of runoff/river discharge (**Q**), **potential evapotranspiration** (**E**) and soil moisture and groundwater storage (**S**) can be expressed as an equation:

$$P = Q + E +/- S$$

These factors can be linked to an annual budget graph such as Figure 2 on page 16 of the student book.

Rainfall effectiveness and water availability

During the summer, potential evapotranspiration is greater than precipitation (Figure 2). So water levels in rivers, lakes and ponds are lower and **soil moisture** levels reduce.

Climate graphs for two different areas can be found on page 17 of the student book. They have similar total rainfall but very different rainfall effectiveness.

- In the UK, the 'water year' begins in October when rainfall generally exceeds evaporation and storage areas are **recharged**.
- Overland flow increases until there is a **surplus**, with the risk of flooding.
- Rising summer temperatures and reduced rainfall result in a **deficit**. Any rainfall remaining after evaporation is **effective rainfall**.

	Jan	Feb	Mar	Apr	May	June	July	Aug	Sept	Oct	Nov	Dec
Rainfall (mm)	78	59	48	49	54	48	58	60	63	70	75	77
Potential evapotranspiration (mm)	5	11	31	50	82	95	97	88	60	33	12	5

🔺 **Figure 2** *Water budget data for south Hampshire*

Soil moisture in the hydrological cycle

The ability of soil to hold moisture is vital to the hydrological cycle. **Field (infiltration) capacity** is the maximum a soil can hold. This capacity is reduced during periods of **water deficit** – when evapotranspiration exceeds precipitation.

Water surplus is when soils are saturated above field capacity, creating surface runoff. **Flash floods** often occur during periods of heavy rainfall after a period of drought when soils cannot absorb the water quickly enough.

Patterns of flow

Factors determining a river's flow (**discharge**) include:

- characteristics of the drainage basin (shape, geology, soil type, land cover)
- inputs and outputs
- climate
- human intervention.

The annual pattern of flow is known as a river's **regime**.

 A comparison of three complex river regimes can be found on page 19 of the student book.

Simple regime

This occurs where a river experiences seasonal high discharge followed by low discharge. This is typical of many rivers in temperate climates where snowmelt occurs (e.g. Rhône) or that are dependent on seasonal storms (monsoons).

Complex regime

This occurs where a river crosses several relief and climatic zones and can be affected by different climatic events (e.g. Ganges). Human factors such as damming or irrigation can increase complexity.

Storm hydrographs

A **storm hydrograph** shows the effects of an individual storm. It shows:

- discharge at a point over a period of time
- the lag in time between the start of a storm and an increase in discharge. This is because most rain falls on the surrounding landscape and takes time to reach the river.

 A storm hydrograph can be found on page 19 of the student book. The factors affecting how and why storm hydrographs vary are on page 20.

Players in the hydrological cycle: planners

UK planners decide whether a proposed development will affect flood risk. Environmental factors (maintaining existing river flow) need to be weighed against economic development (which changes river flow) (Figure 3).

Sustainable Drainage Systems (SuDS)

SuDS reduce runoff produced from rainfall by using:

- green roofs
- infiltration basins
- permeable pavements
- rainwater harvesting
- soak-away
- filter drains
- detention basins
- wetlands.

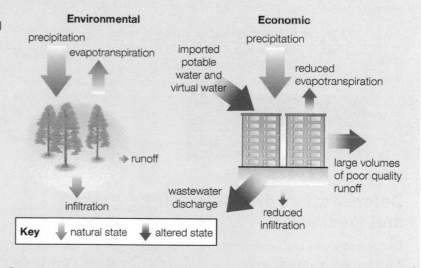

 Figure 3 Environmental factors versus economic development

Ten-second summary

- The water budget shows the balance between inputs and outputs in a given area.
- The capacity of soil to retain moisture is crucial to the hydrological cycle.
- A storm hydrograph shows how a river responds to a storm.
- Planners assess potential changes to flood risk.

Over to you

Create a spider diagram on 'Why river discharge fluctuates'. Expand each 'arm' as far as possible to add greater detail.

You need to know:

- the causes of drought
- how human activity adds to the risk of drought
- the impacts of drought on ecosystems and their resilience.

Big idea

Deficits within the hydrological cycle result from physical processes but can have significant impacts.

Drought in Brazil, 2014–15

Droughts exist when there is a **water deficit** compared to the average rainfall for the same period. In Brazil, rainfall is usually predictable (Figure 1).

- Moist air moves westwards from the South Atlantic across the Amazon Basin.
- The moist air turns southwards at the Andes mountains, maintaining the flow of moisture around the Basin.

In 2014–15, Brazil witnessed the worst drought for 80 years. High-pressure systems diverted rain-bearing winds north, so heavy rain fell in Bolivia and Paraguay, not Brazil (Figure 1). Impacts of the drought were:

- water rationing
- hydroelectric power production ceased, with subsequent power cuts
- reservoirs dried up, some down to just 1% of capacity
- increased groundwater abstraction, reducing aquifers to dangerously low levels
- a reduced crop of Arabica coffee beans, increasing global coffee prices by 50%.

> **Figure 1** *A high-pressure system over the Amazon Basin brought drought to Brazil and heavy rain to surrounding countries*

More detail on this drought can be found on page 22 of the student book.

Types of drought

There are three types of drought:

- **Meteorological** – reduced precipitation compared to 'normal'.
- **Agricultural** – insufficient water for irrigation of crops.
- **Hydrological** – drainage basins suffer shortfalls.

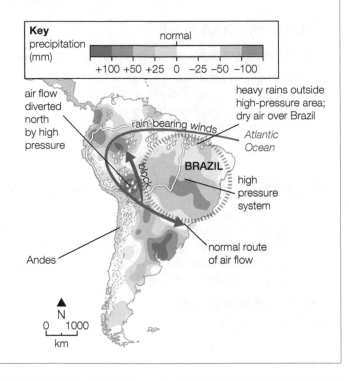

Deforestation, droughts and feedback

Deforestation in the Amazon may have passed a **tipping point**, changing hydrological and climatic cycles permanently. The **positive feedback loop** of deforestation (and less rainfall) reduces the ability of the rainforest to regenerate and to recycle rainfall (Figure 2).

Tropical forests are important in regulating regional climate and generating flows of moisture across the continent. But global climate change, ENSO and deforestation may change this flow so that extreme weather becomes more frequent.

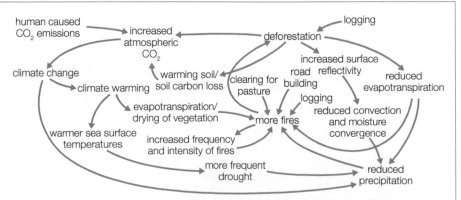

> **Figure 2** *The knock-on effects of feedback loops and the 'domino effect', which are drying out tropical rainforests*

 Find more on the importance of the Tropics on page 10 of the student book and more about ENSO on page 33.

Human activity and drought

Human activity (over-abstraction) contributed to Brazil's 2014–15 drought. Residents tried to avoid cuts in supply by illegal drilling (licenses for water wells are expensive). It is estimated that 70% of all wells were illegally drilled. This created further issues of water contamination.

The impact of drought on rainforest ecosystems

The Amazon rainforest is referred to as 'the Earth's lungs'. It absorbs carbon dioxide and returns oxygen to the atmosphere. Prolonged drought causes forest stress and can cause a series of chain reactions (Figure 3).

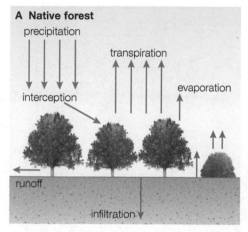

A Native forest
precipitation
transpiration
evaporation
interception
runoff
infiltration

B Following forest die-off
precipitation
transpiration
evaporation
interception
runoff
infiltration

- Younger trees die, reducing the canopy cover.
- This reduces humidity, water vapour and therefore rainfall.
- Dying vegetation and surface tree litter can easily catch fire.
- High winds often turn small fires into wildfires.

🔺 **Figure 3** *Changes in water **fluxes** as forests die due to drought (the number and length of arrows show the effectiveness of each flux)*

The impacts of drought on wetland ecosystems

Wildlife of the Pantanal in South America depends on the permanent wetland of the floodplain for survival. Usually, rainfall floods the Pantanal between November and April (Figure 4), changing the area into an aquatic habitat. The areas near the river are forested, with savanna grassland further away. These areas were most affected by the 2014–15 drought.

- Trees died at an increased rate, reducing animal habitats, cattle ranching and ecotourism.
- Ranchers cleared ungrazed land by burning, but fires spread to the wetlands and forests.

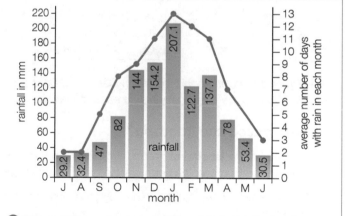

🔺 **Figure 4** *The average monthly rainfall distribution for Corumba, Brazil, in the Upper Paraguay River Basin*

⏱ **Ten-second summary**

- Water deficit over a prolonged period of time causes drought.
- Changes to weather systems led to Brazil's 2014–15 drought.
- There are meteorological, agricultural and hydrological droughts.
- Human activities contribute to drought.
- Drought in rainforests causes dying vegetation, thinner canopies and increased wildfire risk.
- Drought in wetlands causes trees to die, reduces habitats and increases wildfire risk.

✏ **Over to you**

1 In a table with two columns, compare the normal pattern of rainfall in Brazil with what happened in 2014–15.
2 Explain how droughts in the Amazon region illustrate a positive feedback loop.

You need to know:

- the meteorological causes of flooding
- that human actions can exacerbate flood risk
- how damage from flooding has both environmental and socio-economic impacts.

Big idea

Surpluses within the hydrological cycle can lead to flooding, with significant impacts for people.

A recurring deluge – Cumbria, 2015

Places in Cumbria are the wettest in England. Yet flooding in 2005, 2009 and 2015 was extreme, which suggests that the local hydrological system is changing.

The 2015 floods in Cumbria were a result of Storm Desmond, caused by a deep Atlantic **low-pressure** system (depression) (Figure 1). 341.5 mm of rain fell at Honister Pass in 24 hours, for example. It caused extensive flooding (Figure 2).

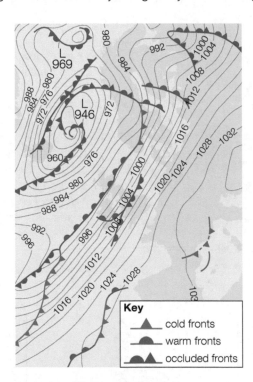

Key
- cold fronts
- warm fronts
- occluded fronts

⬥ **Figure 1** *Synoptic chart for Storm Desmond, December 2015, showing the deep depression east of Iceland*

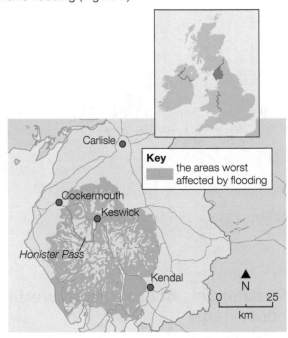

Key
the areas worst affected by flooding

⬥ **Figure 2** *Worst areas of flooding in Cumbria, December 2015*

What caused the flooding?

The position of the **jet stream** determines the direction and speed of depressions. In 2015, it remained over the north-west longer than usual, bringing Storm Desmond.

- Warm, moist air was forced to rise by the Cumbrian fells, creating orographic (relief) rainfall.
- The moist air mass sat over Cumbria for up to 48 hours, with record rainfall and **flash flooding**.
- The saturated ground, impermeable surfaces and blocked drains caused overland flow and flooding, especially at river confluences at Cockermouth and Carlisle.

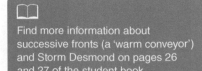

Find more information about successive fronts (a 'warm conveyor') and Storm Desmond on pages 26 and 27 of the student book.

Other causes of flooding

- **Storms and flash flooding** – often associated in the UK with extreme rainfall in summer, when river capacity cannot cope.
- **Monsoons** – in South and South-East Asia between May and September, e.g. in 2016, a larger than normal

low-pressure system caused flooding, landslides and evacuation from villages in the Philippines.

- **Snowmelt** – causes flooding when it cannot infiltrate into already saturated soil, e.g. Norfolk 2013.

Exacerbating flood risk

Changing land use

Landscape management impacts on flood risk.

- Overgrazing of previously forested areas can result in bare slopes, faster runoff, reduced stream lag times and higher peak discharge. Trees previously absorbed water and released it slowly.
- Straightened, dredged channels mean rainwater now reaches floodplains quicker.
- As urban areas expand, floodplains have increasingly impermeable surfaces.

Mismanaging rivers

Hard-engineering schemes are common in Cumbria. But, with more extreme storms (climate change) and further changes in land use, small-scale schemes may be overwhelmed by flooding more frequently. Higher rainfall totals mean flooding may now be inevitable.

Mitigating flood risk

The Environment Agency now believes that **soft-engineering** schemes should be used rather than **hard-engineering** schemes, which are expensive and are not effective in extreme cases. Examples of soft engineering are:

- reafforestation
- restoration of meandering river channels
- restoration of absorbent floodplains to store floodwater
- discouraging building near rivers.

Longer-term impacts of flooding

The cumulative effect of three flood events in quick succession (2005, 2009, 2015) forced key **players** in Cumbria to weigh up the costs and benefits of staying where they are (Figure 3). Flood risk inevitably affects the long-term plans of investors, residents and businesses.

Social costs	<ul><li>3000 homes flooded in 2005, over 5200 in 2015.</li><li>Residents of flooded properties had to live in temporary accommodation.</li><li>Some local services, such as schools, healthcare, shops and offices were forced to close temporarily.</li><li>Many residents suffered anxiety, stress and psychological trauma.</li></ul>
Economic costs	<ul><li>Many businesses closed, and transport links and infrastructure were damaged.</li><li>The costs in Cumbria were £100 million in 2005, £270 million in 2009 and £400–500 million in 2015.</li><li>Insurance claims caused by flooding across the UK in 2015 exceeded £6 billion.</li><li>Farmers lost hedgerows and expensive dry-stone walls, and many sheep drowned.</li><li>House prices fell in flood-risk areas.</li><li>The risk of repeated flooding deterred tourists from visiting.</li></ul>
Environmental costs	<ul><li>Many river banks were eroded, adding to future flood risk.</li><li>Rivers were choked with debris and contaminated with sewage and pollutants.</li><li>Soils were eroded, habitats destroyed and ecosystems affected.</li><li>Decomposition of dead plants and animals gave off noxious gases. Other poisons contaminated the food chain and threatened wildlife.</li><li>Saturated ground led to landslides.</li></ul>

Figure 3 *The social, economic and environmental costs of flooding in Cumbria, 2015*

Ten-second summary

- In 2015, Storm Desmond caused widespread flooding and disruption in Cumbria.
- Meteorological causes of flooding include low-pressure systems (depressions), the position of the jet stream, intense storms, monsoon rainfall, snowmelt.
- Human actions can increase the risk of flooding.
- Soft-engineering solutions may mitigate flood risk.
- There are longer-term social, economic and environmental costs of flooding.

Over to you

1 List:
   a the meteorological causes of flooding
   b ways in which human actions exacerbate flood risk.
2 Write down four of each of the social, economic and environmental impacts of the 2015 floods.

You need to know:

- that climate change affects the hydrological cycle
- that climate change affects stores and flows
- that ENSO cycles and global warming cause uncertainty over the security of water supplies.

Big idea

Climate change may have significant impacts on the hydrological cycle both globally and locally.

Good, bad and unpredictable

The climate is changing. Globally, more rain in some places might reduce current water scarcity, but a deficit could result in other areas suffering drought.

Droughts will mean:

- reduced inputs
- reduced storage in soil, rivers and lakes
- groundwater flow becoming more important
- high rates of evaporation.

The good – the Sahel

The Sahel forms the southern fringe of the Sahara Desert (Figure 1).

- Annual rainfall varies from 100 mm to 600 mm. Figure 2 shows variations in rainfall.
- Most rain falls during the monsoon from July to September.
- The monsoon rains failed in the 1970s and 1980s, causing drought.
- Since 1996, there have been several wet years.
- **Re-greening** to create productive farmland may be a possibility.

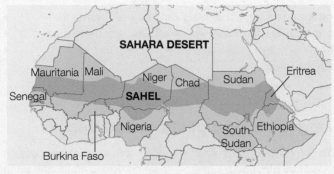

▲ **Figure 1** *The Sahel region and the countries it runs through*

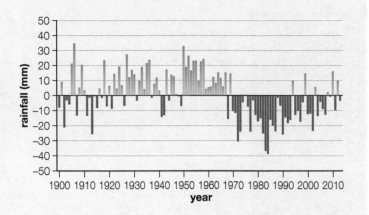

▶ **Figure 2** *Rainfall variations in the Sahel during each rainy season in comparison with average rainfall for 1900–2013 (at '0')*

The bad – California

California is facing increasing water problems because of variations in rainfall. Evidence points towards climate change, with decade-long dry periods a distinct possibility.

2015 saw the fourth year of drought due to rising temperatures (with subsequent **increased evaporation rates**) and reduced precipitation. The consequences of this are:

- water rationing
- farmers abandoning their fields
- increased risk of wildfires.

There is already evidence of problems in meeting water demand, such as:

- Surface runoff and soil moisture levels have declined.
- Forests have become scrub and grassland.
- Groundwater levels fell by 30 metres between 2011 and 2015.
- On average, reservoir levels have fallen.
- Permanent snow levels were at a record low in 2015; meltwater is crucial in the water supply.

Global climate patterns

Global climate patterns are mainly caused by atmospheric circulation. Ocean currents also play a part, as surface winds take surface water with them. Figure 3 summarises the normal situation of winds circulating around the **Walker Cell** in the Pacific Ocean, and also during **La Niña** and **El Niño**.

See page 32 of the student book for more detail on La Niña and El Niño.

	West Pacific pressure	West Pacific rainfall	Direction of surface flow	East Pacific pressure	East Pacific rainfall
Normal conditions Winds circulate around the Walker Cell	Low	Heavy	East–West	High	Low/none (dry)
La Niña (normal conditions intensified) Occurs just before or just after El Niño	Intense low	Very heavy	East–West	Intense high	Drought
El Niño Occurs every 3–8 years and can last 14–22 months	High	Dry	West–East	Low	Heavy

Figure 3 *Normal conditions around the Walker Cell, and during La Niña and El Niño*

Predicting future climate change

The change in air pressure between 'normal' years and El Niño is the **El Niño Southern Oscillation (ENSO)**. It is the difference in pressure measured between Easter Island (west of South America) and Darwin (Northern Australia). A sharp drop in pressure means El Niño is imminent.

Futures and uncertainties

El Niño events have been happening for around 15 000 years, but climate change may be increasing their duration and intensity. This could have a significant impact on water supplies. Predictions are uncertain (Figure 4), but as the planet warms:

- different parts of the world warm at different rates
- La Niña and El Niño events will become more frequent and intense
- Pacific regions will have floods following drought, and vice versa.

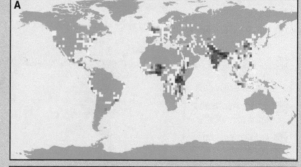

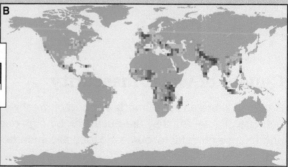

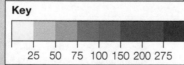

Key

| 25 | 50 | 75 | 100 | 150 | 200 | 275 |

Figure 4 *Projected changes in people's exposure to floods (**A**) and drought (**B**) by 2100. The dark colours show greatest projected increases in exposure*

Ten-second summary

- Climate change affects the inputs, stores, flows and outputs of the hydrological cycle.
- Negative effects include increased droughts in California.
- La Niña and El Niño are fluctuations in the global climate cycle that affect global weather patterns.
- Climate change may be increasing the characteristics of ENSO events, which could impact on water security.

Over to you

1 List three ways in which climate change affects stores and flows.
2 Draw two simple sketch maps of the Pacific Ocean to show:
   a a La Niña event   **b** an El Niño event.
   Include South-East Asia, Australia and western South America. Annotate the location of low/high pressure areas, wet/dry areas and the direction of the winds for each event.

Student Book
See pages 34–37

You need to know:

- that increasing demand has led to global water stress and water scarcity
- about the physical and human causes of water insecurity
- that water resources are finite and rising demand is leading to serious regional water insecurity.

Big idea

There are physical and human causes of water insecurity.

A global water crisis

In theory, there is no global water shortage. Only 50% of available water is actually used. However:

- '*The lack of freshwater is emerging as the biggest challenge of the twenty-first century*' (UN) (Figure 1).
- Water shortages and a lack of safe water affect a third of the world's population.
- Rapid population growth in areas of limited supplies, uneven global distribution of water and deterioration in water quality all increase the number of people facing severe water shortages.

Water scarcity, stress and insecurity

Water scarcity: less than 1000 m³ available per person per year.
Physical scarcity: not enough water to meet demand.
Economic scarcity: water is available but people cannot afford it.
Water stress: less than 1700 m³ available per person per year.
Water insecurity: present and future supplies cannot be guaranteed.

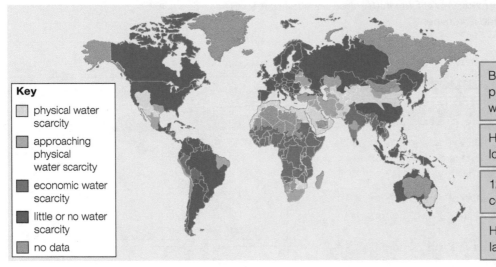

◀ **Figure 1** *Significant problems caused by water scarcity*

Key
- physical water scarcity
- approaching physical water scarcity
- economic water scarcity
- little or no water scarcity
- no data

By 2025, half the world's population will be living in water-deficit areas.

Half the world's rivers no longer flow all year.

12% of the world's population consumes 85% of its water.

Half the world's rivers and lakes are badly polluted.

Causes of water insecurity

Health, welfare and livelihoods depend on freshwater being available. But they are put at risk by high demand and misuse of water resources, especially in certain regions (Figure 2).

Groundwater contamination, Puerto Rico 2015

In 2015, seepage from a leaking industrial tank at the Retiro Industrial Park in San German, Puerto Rico contaminated the wells supplying public water, together with the soil and groundwater stores.

For more detail on reasons for regional water insecurity, see Figure 3 on page 35 of the student book.

Region	Water issues
Asia and the Pacific	• Lack of clean water causes many deaths by diarrhoea • Bacterial waste in water is often up to 10 times greater than safe levels • Overuse and resource depletion • Raw sewage and industrial pollutants are dumped into rivers
Europe/ Central Asia	• Lack of access to clean water (Eastern Europe/Central Asia) • Increasing consumption • Groundwater contamination
North America	• Aquifers are being depleted • Climate change is affecting rainfall patterns (La Niña/El Niño) • Pollution from runoff
Africa	• 19 of the 25 countries with lowest access to clean water are in Africa • Groundwater is being depleted by agricultural use • Floods and droughts displace huge numbers of people
Latin America/ Caribbean	• Groundwater contamination and depletion (e.g. Puerto Rico) • Only 2% of all sewage is treated • Economic scarcity

▲ **Figure 2** *Causes of regional water insecurity*

Under pressure

In the 20th century, global population increased four times but water consumption increased six times. The rise in water consumption was due to improved living standards and increased access to water (Figure 3).

Increasing population and urbanisation	• Demand is growing twice as fast as population • Half the world's population live in urban areas. By 2030, the urban population in Africa and Asia is expected to double
Improving living standards	• Rising incomes in developing and emerging economies • Increased meat consumption • Larger homes • Production of more cars, appliances and gadgets
Industrialisation	• Industrial consumption will increase by 400% by 2050 • Increased industrial pollution contaminates supplies
Agriculture	• Largest consumer of water • By 2050, 60% more food will be required • Depleting aquifer levels and contamination by salt water

Figure 3 *Reasons for increased global water scarcity*

Futures and uncertainties about water stress

High water stress means that some countries are vulnerable to variations in supply. This threatens not only water security but also economic growth. Climate change will make some areas drier. Figure 4 shows that:

- 33 countries are predicted to face high levels of water stress by 2040.
- 14 of these are in the Middle East, where groundwater and desalination are crucial.
- The USA, China and India face insecurities, and Chile is likely to be highly stressed by 2040.

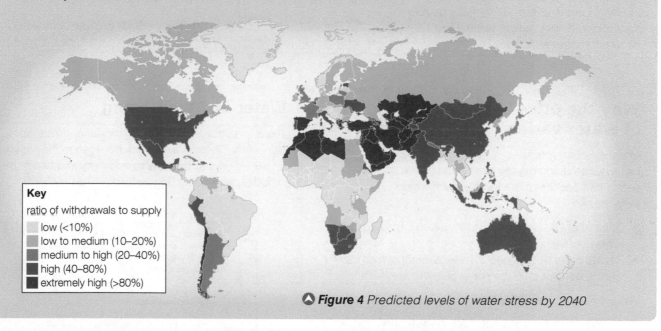

Key

ratio of withdrawals to supply

- low (<10%)
- low to medium (10–20%)
- medium to high (20–40%)
- high (40–80%)
- extremely high (>80%)

Figure 4 *Predicted levels of water stress by 2040*

 Ten-second summary

- Demand outstripping supply is leading to a global pattern of water stress and scarcity.
- There are physical and human causes of water insecurity.
- There is increased demand from a growing population, increased urbanisation, improved living standards, industrialisation and agriculture.
- The number of countries with high water stress is increasing, threatening security and economic growth.

 Over to you

1 Define the terms 'water stress' and 'water scarcity'.
2 List the reasons for water insecurity. Then write one human and one physical reason on separate sticky notes. Stick them up in the house to help you remember.

Student Book
See pages 38–43

You need to know:

- the causes of water scarcity
- why the price of water varies globally
- the importance of water supply for economic development and human well-being
- the problems resulting from inadequate water supply, including conflict.

Big idea

There are consequences and risks associated with water insecurity.

Causes of water scarcity

There are two kinds of water scarcity:

Physical water scarcity

Factors causing physical scarcity include:

- low rainfall and high temperatures
- climate change
- human activity affecting the availability of resources.

Economic water scarcity

The main factors causing economic scarcity are the high cost of collection, storage, purifying and distribution, so many people cannot afford 'processed' water. Figure 1 shows comparative costs.

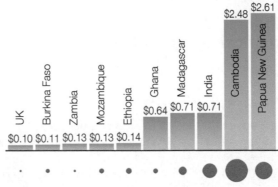

UK	Burkina Faso	Zambia	Mozambique	Ethiopia	Ghana	Madagascar	India	Cambodia	Papua New Guinea
$0.10	$0.11	$0.13	$0.13	$0.14	$0.64	$0.71	$0.71	$2.48	$2.61
0.1%	9%	4%	13%	15%	13%	25%	45%	108%	54%

▲ **Figure 1** *Cost of 50 litres of water (in US$) in LICs, compared with the UK (including % of typical low-income worker's daily wage)*

Why the price of water varies

See page 39 in the student book for information about the Water Poverty Index.

Massive urban growth has meant that the private sector has become more involved with water services. So:

- Consumers have to pay for improvements to ageing infrastructure.
- Private companies need to make a profit, which can lead to conflict (see the panel below).

Water charges vary in HICs.

- Canada's water is still publically owned and consumers pay 80% less than in Germany.
- Ireland's water bills are 75% less than in the UK (Ireland only started charging in 2012).
- Denmark's water is expensive as the price is used to try to cut consumption.

Water privatisation in Bolivia, 1999

In 1999, the water system in Cochabamba, Bolivia was privatised. The price of water supplies was increased to 20% of the average income of Cochabamba's urban poor. Four days of protests led to the Bolivian government cancelling the contract.

Water and well-being

There is a strong link between wealth and access to safe water. There is also a link between money spent on sanitation and benefits to economy and health, including well-being (Figure 2).

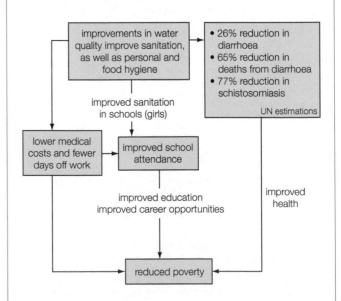

▲ **Figure 2** *How sanitation can lead to increased health, wealth and well-being*

Water and economic development

Water is essential for producing goods and services. And, as an economy develops, demand for water increases. A 40% **water gap** (shortfall) in supply could exist by 2030.

Global demand is increasing because:

- **Energy generation** needs water – 75% of UK water consumption is linked to this.
- Global food production – **agriculture** needs to increase by 60% by 2050.
- Use of water is not always sustainable so efficiency needs to be increased.

Water and potential for conflict

More risk of water shortages means more potential for conflict.

- 263 rivers form or cross international boundaries and 90% of countries share a drainage basin with another country.
- Dams and diversion schemes, built to meet increasing demand, can adversely affect river flows.

See page 41 of the student book for examples of contested water resources, including those shared by Turkey, Syria and Iraq.

The Murray–Darling Basin (MDB)

The MDB provides 75% of Australia's water and 40% of its farm produce. Increasing demand and mismanagement have put its supplies under threat.

- There has been a five-fold increase in water extraction in the MDB since the 1920s.
- The MDB crosses several natural environments, so there are wide variations in rainfall.
- Water extraction needs to be well managed to make sure all areas are supplied with water.

In 2012, a new agreement set limits on the amounts of water used by **agriculture**, **industry**, **urban** residents and the **government**. The aim was to make sure that there was water for all, with no negative impact on the natural environment.

A success?

Rural communities in the MDB claimed that not enough water was being made available for **irrigation**, so farmers could no longer grow food and communities were dying. In one area, 500 farm jobs were lost between 2012 and 2014, and the population fell by 18%. Yet farmers complained about 'environmental' water flowing freely down the rivers.

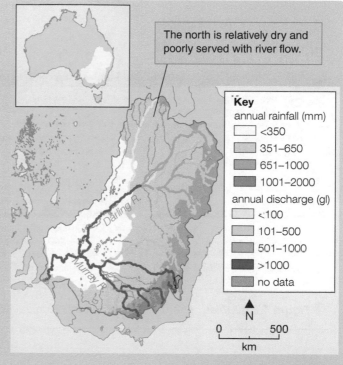

The north is relatively dry and poorly served with river flow.

Key

annual rainfall (mm)
- <350
- 351–650
- 651–1000
- 1001–2000

annual discharge (gl)
- <100
- 101–500
- 501–1000
- >1000
- no data

N

0 — 500 km

 Figure 3 *Average rainfall and river flow in the MDB*

See page 42 of the student book for key players in the MDB.

 Over to you

1 Draw two spider diagrams:
  a to show the reasons why the price of water varies globally
  b to explain why water is essential for development.
2 Give three reasons why water insecurity, or the fear of water insecurity, could lead to conflict.

You need to know:

- the pros and cons of hard-engineering schemes
- the value of managing water supplies sustainably
- about integrated drainage basin management and water-sharing treaties.

Big idea

There are different approaches to managing water supply, some more sustainable than others.

Managing water supplies – hard engineering

Hard engineering, such as dams, water transfer projects and desalination, are long-standing methods of managing water supplies (Figure 1).

Key facts	Pros	Cons
Water transfer (South–North Water Transfer Project, China)		
• The Beijing region has only 7% of China's water. • It will take water from the Yangtze to northern China. • The cost is US$70 billion. • It is due for completion by 2050.	• It will reduce the risk of water shortages in Beijing and boost economic development. • It will reduce the abstraction of groundwater.	• It will submerge 370 km² of land. • 345 000 people will have to relocate. • It risks draining too much water from southern China. • The Eastern route is industrial and risks further pollution.
Mega dam (Three Gorges Dam, China)		

• It was designed to:
  – control flooding on the Yangtze
  – improve water supply by regulating river flow
  – generate HEP
  – make the river navigable.
	• It enables surplus water to build up and be diverted to northern China via the South-North Water Transfer Project.
	• The electricity generated is vital for China's growth.
		• It flooded 632 km² of land.
		• 1.3 million people from 1500 villages and towns were relocated.
		• Water quality is low because of industrial pollution, farm waste and sewage.
		• Decomposing vegetation produces methane in the reservoir.
		• It was controversial and very expensive.

Desalination (Israel)		
• Five plants take water directly from the Mediterranean Sea. • The aim is to provide 70% of Israel's domestic water supplies by 2020.	• Desalination plants provide a reliable and predictable supply of water. • They produce up to 600 tonnes of potable water per hour.	• Each plant requires its own power station and adds to CO_2 emissions, but much of the energy used is solar. • They produce vast amounts of salt/brine, which harm ecosystems.

▲ **Figure 1** *Examples of hard-engineering technological fixes*

Managing water supplies sustainably

Water management in Israel

Israel needs to manage its water supply efficiently because of its climate, natural geography and politics. The National Water Carrier is a north–south water transfer project. Israel also manages limited supplies through conservation methods such as:

- **smart irrigation**
- **importing virtual water**
- **recycling**
- charging **'real value' prices**
- **reducing** agricultural **consumption**.

It also obtains new supplies by importing water from Turkey and desalinating water from the Red Sea and Mediterranean.

Holistic management in Singapore

Malaysia has traditionally supplied 80% of Singapore's water. By 2010, the volume had halved, due to Singapore's careful management of water, including:

- water metering and education
- cutting leakage
- higher charges for higher use
- subsidies for the poor
- water harvesting, recycling, desalination.

Restoring aquifers in Saudi Arabia

In the 1980s, Saudi Arabia increased wheat production by extensive use of irrigation. As a result, aquifer levels fell significantly. To reduce demand on aquifers, grain is now imported and wheat production is virtually zero (Figure 2).

Production was reduced annually by regulation from 2007, and imports increased.

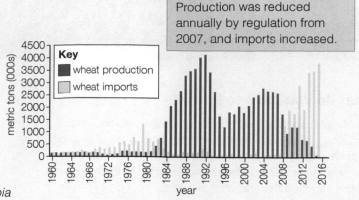

Key
■ wheat production
□ wheat imports

 Figure 2 *Changes to wheat supply in Saudi Arabia*

Integrated drainage basin management – sharing the Colorado

For most of the 20th century, water has been distributed from the Colorado to seven US states and Mexico. The 1922 Colorado Compact agreement is now out of date due to population growth, economic development and lower rainfall. New agreements have now updated it.

- **2007 agreement** – determined that supplies to each state depended on the water level of Lake Mead. Different levels triggered different quantities of water.
- **2012 agreement** – adjusted the amount of water Mexico receives from the Colorado in times of drought or surplus. During periods of surplus, US water providers can buy part of Mexico's allocation.

See page 46 of the student book for background to the issues surrounding water supply from the Colorado.

Changing attitudes and actions to water supply

Water is seen as an entitlement in more developed areas and this can lead to conflict. This can be avoided by:

- conservation of domestic supplies
- reusing wastewater
- saving storm water
- reducing irrigation
- smart planning.

See page 47 of the student book for the players in reducing water conflict and more on water-sharing treaties and frameworks.

 Ten-second summary

- There are pros and cons to hard-engineering approaches to water management.
- There is a variety of sustainable ways of managing water supplies.
- Integrated drainage basin management agreements aim to ensure water is shared fairly.
- Conservation is being encouraged and education teaches that water is not an entitlement.
- Intergovernmental organisations (IGOs) and governments work to improve the management of water resources and reduce the risks of conflict.

Over to you

Without looking back at the text, write down two pros and two cons about maintaining water supplies by each of the following:

a water transfer projects
b dams
c desalination.

Include examples.

Chapter 2
The carbon cycle and energy security

What do you have to know?

This chapter studies the importance of the carbon cycle in maintaining planetary health and how it operates at a range of spatial scales and timescales. Physical processes control the movement of carbon between stores on land, the oceans and the atmosphere, and changes to the carbon cycle are a result of physical and human processes. Reliance on fossil fuels has caused significant changes to carbon stores and contributed to climate change.

The specification is framed around three enquiry questions:

1 How does the carbon cycle operate to maintain planetary health?
2 What are the consequences for people and the environment of our increasing demand for energy?
3 How are the carbon and water cycles linked to the global climate system?

The table below should help you.

- Get to know the key ideas. They are important because 12- and 20-mark questions are likely to be based on these.
- Copy the table and complete the key words and phrases by looking at Topic 6 in the specification. Section 6.1 has been done for you.

Key idea	Key words and phrases you need to know
6.1 Most global carbon is locked in terrestrial stores as part of the long-term geological cycle.	biogeochemical carbon cycle, carbon stores (terrestrial, oceans and atmosphere), fluxes, sedimentary carbonate rocks, biologically-derived carbon, geological processes, volcanic out-gassing
6.2 Biological processes sequester carbon on land and in the oceans on shorter timescales.	
6.3 A balanced carbon cycle is important in sustaining other Earth systems but is increasingly altered by human activities.	
6.4 Energy security is a key goal for countries, with most relying on fossil fuels.	
6.5 Reliance on fossil fuels to drive economic development is still the global norm.	
6.6 There are alternatives to fossil fuels but each has costs and benefits.	
6.7 Biological carbon cycles and the water cycle are threatened by human activity.	
6.8 There are implications for human wellbeing from the degradation of the water and carbon cycles.	
6.9 Further planetary warming risks large-scale release of stored carbon, requiring responses from different players at different scales.	

You need to know:

- that burning fossil fuels affects climate
- that the use of fossil fuels drives economic development
- why the price of oil fluctuates.

Big idea

There is a conflict between increasing energy use and maintaining the global climate.

The conflict

In December 2015, 195 countries adopted the first legally binding global climate deal at the Paris Climate Conference (COP21).

- Governments agreed to a long-term goal of keeping the increase in global temperature below 2°C above pre-industrial levels.
- This requires serious reductions in global greenhouse gas emissions.
- The problem is that the world relies on fossil fuels and burning them releases carbon as carbon dioxide (CO_2), the main driver of climate change (Figure 1).

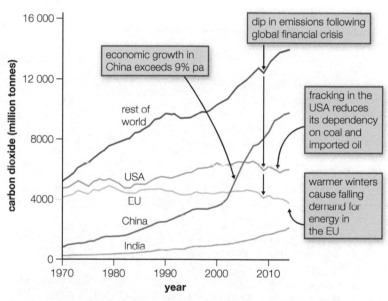

▲ **Figure 1** Carbon dioxide emissions from energy use, 1970–2015

Giving up fossil fuels? The case of India

For some countries, abandoning fossil fuels could threaten economic development. India, for example:

- depends on coal for 66% of its energy and intends to double its coal output by 2020
- is the third largest CO_2 emitter after China and the USA, and wants to reduce its dependence on imported fuel
- but this wish means using more domestic coal.

New infrastructure, an expanding middle class and 600 million new users of electricity are driving India's demand for coal.

Why do oil prices fluctuate?

The price of oil reflects political and economic factors, including market demand.

- In the past, if demand rose, OPEC producers tended to increase oil production to prevent sharp price increases or to reduce production to maintain price if demand fell.
- Fracking in North America has changed this. The USA is the world's largest oil consumer and drives global prices. US oil prices have fallen sharply since 2012 (Figure 2), due to huge new supplies of oil and shale gas from the USA and Canada.
- OPEC has cut prices to compete, to maintain its market share and for geo-political motives.

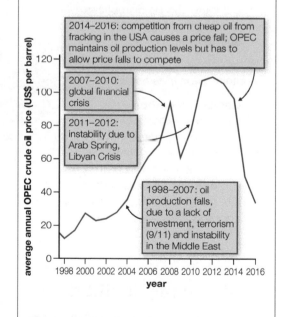

▲ **Figure 2** Changes in the OPEC crude oil price, 1998–2016

 Ten-second summary

- Burning fossil fuels releases CO_2, the main driver of climate change.
- Abandoning fossil fuels could threaten economic development in some countries.
- Oil price reflects political and economic factors.

Over to you

Explain the links between fossil fuel consumption, climate change and oil prices.

You need to know:

- about the geological and bio-geochemical carbon cycles
- that carbon is either geologically or biologically derived
- the processes that help to maintain equilibrium in the carbon cycle.

Big idea

Global carbon is locked in terrestrial stores as part of the long-term geological cycle.

Understanding carbon

Carbon provides the building blocks for life on Earth. It regulates climate and is **stored** within rocks, plants and oceans.

- **Stores** are also known as **pools, stocks** and **reservoirs**.
- There are **terrestrial** (land), **oceanic** and **atmospheric** stores.
- **Flux** refers to the **movement/transfer** of carbon between stores (e.g. land and sea). Fluxes create cycles and feedbacks.

The geological carbon cycle

The geological carbon cycle (Figure 1) transfers carbon between land, oceans and atmosphere.

- There tends to be a balance between carbon production and absorption within this cycle.
- But there can be disruptions and short periods before **equilibrium** is restored.

The geological carbon cycle contains two types of carbon.

- **Geological carbon** (i.e. rocks) results from the formation of sedimentary carbonate rocks (limestone and chalk) in oceans.
- **Biologically derived carbon** (i.e. from plants) is stored in shale, coal and other sedimentary rocks.

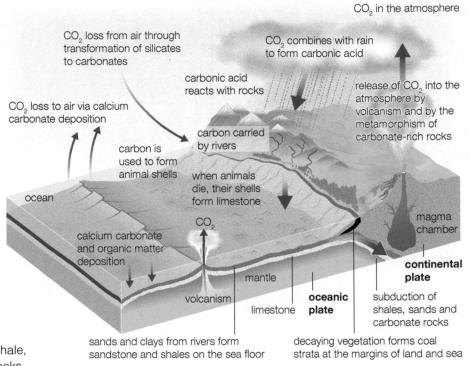

Figure 1 *The geological carbon cycle*

Maintaining equilibrium

Equilibrium between stores is maintained, though slowly.

- **Outgassing** occurs when terrestrial carbon within the mantle is released **into** the atmosphere as CO_2 when volcanoes erupt.
- **Chemical weathering** occurs when CO_2 in the atmosphere combines with rainfall (taking it **out** of the atmosphere) to produce weak carbonic acid, which dissolves carbon-rich rocks. This releases bicarbonates, which are eventually deposited as carbon on the ocean floor.

The bio-geochemical carbon cycle

Biological and chemical processes determine how much carbon on the Earth's surface is stored or released at any time. So it's called the **bio-geochemical carbon cycle**. Living organisms are critical in maintaining this system because they control the balance between storage, release, transfer and absorption.

Four key processes transfer carbon from one store to another in the cycle (Figure 2):

- **photosynthesis**
- **respiration**
- **decomposition**
- **combustion**.

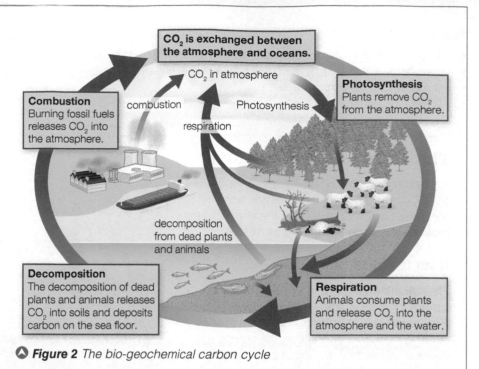

CO₂ is exchanged between the atmosphere and oceans.

CO₂ in atmosphere

Combustion Burning fossil fuels releases CO₂ into the atmosphere.

combustion

Photosynthesis

respiration

Photosynthesis Plants remove CO₂ from the atmosphere.

decomposition from dead plants and animals

Decomposition The decomposition of dead plants and animals releases CO₂ into soils and deposits carbon on the sea floor.

Respiration Animals consume plants and release CO₂ into the atmosphere and the water.

 Figure 2 *The bio-geochemical carbon cycle*

How much carbon is there?

The Earth's total carbon store is very large, but it's the rate of **flux** between stores that matters. It's estimated that the amount of carbon that has been added to the atmosphere from burning fossil fuels is tiny compared to the amounts transferred naturally, but it's still enough to trigger climate change.

 See Figure 4 on page 54 of the student book for the size of carbon stores and fluxes, measured in petagrams (Pg).

Types of carbon

Carbon can be:

- **inorganic** – found in rocks as bicarbonates and carbonate (Earth's largest carbon store)
- **organic** – found in plant material
- **gaseous** – found as CO₂, CH₄ (methane) and CO (carbon monoxide).

Variations in carbon fluxes

Variations over time

- The quickest cycle is completed in seconds, as plants take carbon from the atmosphere through photosynthesis and release it by respiration. Sunlight, temperature and moisture control the speed of these processes.
- Dead organic material in soils may retain carbon for years. Some organic materials may become buried so deeply that they are transformed into sedimentary rocks (e.g. limestone) or hydrocarbons (oil and natural gas). CO₂ is only released when these are burned, or used industrially (e.g. limestone for making cement).

Global variations

CO₂ fluxes vary with latitude. Levels are higher in the Northern Hemisphere because it contains greater landmasses and temperature variations than the Southern Hemisphere.

 Ten-second summary

- The geological carbon cycle moves carbon between land, oceans and the atmosphere.
- Biological and chemical processes transfer carbon between stores within the bio-geochemical carbon cycle.
- Carbon is either geologically or biologically derived.
- Fluxes move carbon between stores, and vary in terms of their speed and by latitude.

Over to you

Draw a spider diagram showing the importance to the carbon cycle of:

- volcanic eruptions
- chemical weathering
- respiration
- decomposition
- photosynthesis
- burning fossil fuels.

You need to know:

- how phytoplankton sequester carbon and how the carbonate pump works
- about the role of terrestrial primary producers in sequestering carbon
- the role of soils as carbon stores.

Big idea

Biological processes sequester (take up) carbon on land and in the oceans.

The biological carbon pump

At the surface of the ocean, there's an exchange of CO_2. Some dissolves in water and some is vented out to the air above. This is known as the **biological carbon pump** (Figure 1).

How it works

Phytoplankton in the ocean's surface layer contain chlorophyll and need sunlight to live. They **sequester** CO_2 through photosynthesis, creating calcium carbonate as their shells develop. When they die, these carbon-rich micro-organisms sink to the ocean floor, accumulating as sediment.

This process is known as the **carbonate pump** and is a part of the biological carbon pump, pumping CO_2 out of the atmosphere and into the ocean store.

It's a fragile system. Phytoplankton require nutrients in vast quantities, and existing ocean temperatures and currents maintain a constant supply.

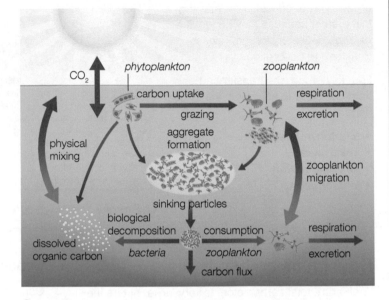

▲ **Figure 1** *The biological carbon pump. In the diagram, zooplankton are animal plankton and phytoplankton are plankton capable of carrying out photosynthesis*

- The recycling of particles that sink in deep waters by upwelling currents is critical.
- The **thermohaline circulation** (see Section 1.2) maintains the pump.
- However, slight changes in water temperature can alter the flow; pollution and turbulence reduce light penetration and slow the pump.
- Research has shown that the Gulf Stream has slowed.
- All these factors are vulnerable to climate change – so the risk of the pump breaking down is real.

Terrestrial stores

Terrestrial (land-based) **primary producers** sequester carbon through photosynthesis. The amount of carbon stored in vegetation and soil varies across the world's biomes.

- In terrestrial ecosystems, carbon is found in plants, animals, soils and micro-organisms (e.g. bacteria and fungi). Leaves, roots, and dead and decaying material in soil all contain carbon.
- Green plants are primary producers that use solar energy to produce biomass – plant growth on land, and algae and phytoplankton in water.

- CO_2 is absorbed and converted into new plant growth during **photosynthesis**.
- Plants release CO_2 into the atmosphere through **respiration**.
- **Primary consumers** (e.g. bugs, larvae, herbivores) feed on producers and return carbon to the atmosphere during respiration.
- Organisms such as insects, worms and bacteria (**biological decomposers**) feed on dead plants, animals and waste.

See Figure 2 on page 57 of the student book for sizes of carbon stores in different biomes.

Mangroves and the role of soil

Biological carbon can be stored in soils as dead organic matter or returned to the atmosphere as a result of decomposition. Deforestation and land use change can release carbon stores very rapidly.

Mangrove forests are found along tropical and sub-tropical tidal coasts in Africa, Australia, Asia and the Americas.
They sequester almost 1.5 metric tonnes of carbon per hectare every year.
Figure 2 shows the carbon cycle in mangrove forests.

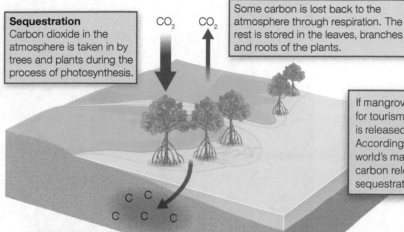

Sequestration
Carbon dioxide in the atmosphere is taken in by trees and plants during the process of photosynthesis.

Some carbon is lost back to the atmosphere through respiration. The rest is stored in the leaves, branches and roots of the plants.

◀ **Figure 2** *The carbon cycle within mangroves*

If mangroves are drained or cleared (e.g. for tourism, shrimp farms, etc.), carbon is released back to the atmosphere. According to researchers, if just 2% of the world's mangroves are lost, the amount of carbon released will be 50 times the natural sequestration rate.

Storage
Dead leaves, branches and roots containing carbon are buried in the soil, which is frequently, if not always, covered with tidal waters. This oxygen-poor environment causes very slow break down of the plant materials, resulting in significant carbon storage (over 10%).

Tundra soils as carbon stores

Much of the soil in the tundra region is permanently frozen.

- Microbe activity only takes place in the surface layer of the soil when it thaws.
- The rest of the time, roots and dead and decayed organic matter are frozen, locking carbon into an icy store.

Tropical forests as carbon stores

Tropical rainforests are huge carbon sinks. Carbon is mainly stored in trees, plant litter and dead wood.

- Litter and dead wood decay and are recycled quickly so a soil store barely develops. Even carbon given off by decomposers is rapidly recycled.
- Tropical rainforests absorb more atmospheric CO_2 than any other terrestrial biome, accounting for 30% of global net primary production.
- Without rainforests, the world would lose a massive carbon sink.

Ten-second summary

- Phytoplankton sequester carbon, die and sink to the ocean floor, accumulating as sediment. This is the carbonate pump.
- The thermohaline circulation maintains the carbonate pump.
- Primary producers sequester carbon during photosynthesis.
- Biological carbon is stored in soils and forests.

Over to you

From memory, write definitions for, and the significance of, each of these terms:

- carbon sequestration
- photosynthesis
- biological carbon pump
- thermohaline circulation
- primary producers
- primary consumers.

You need to know:

- that the concentration of atmospheric carbon influences the greenhouse effect
- how photosynthesis helps to regulate the composition of the atmosphere
- how soil health is influenced by stored carbon
- that fossil fuel combustion has implications for climate, ecosystems and the hydrological cycle.

Big idea

A balanced carbon cycle is important for sustaining other Earth systems but is increasingly altered by human activities.

The natural greenhouse effect

The Earth's energy is received as incoming solar radiation.

- Dark surfaces absorb this and radiate it back as heat.
- CO_2 and CH_4, together with nitrous oxide, halocarbons, ozone and water vapour, (**greenhouse gases**) absorb and reflect some of the radiated heat back to the Earth.
- This keeps the Earth's surface 16°C warmer than it would otherwise be – warm enough to sustain life.

Processes controlling the amounts and concentrations of greenhouse gases in the Earth's atmosphere affect global climate, so CO_2 and CH_4 (carbon-based gases) both contribute to the **natural greenhouse effect** (Figure 1).

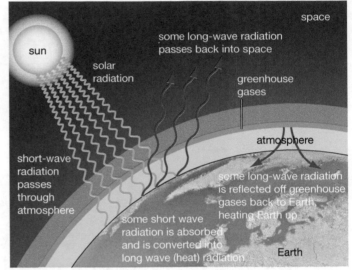

Figure 1 *The natural greenhouse effect*

Greenhouse gases

Greenhouse gases act as a blanket, retaining heat (Figure 2).

Greenhouse gas	% of greenhouse gases produced	Sources	Warming power compared to CO_2	% increase since 1850
CO_2 (has the highest **radiative forcing effect (RFE)** of all gases emitted by human activities, holding onto more heat for longer)	89%	burning fossil fuels; deforestation	1	+ 30%
Methane	7%	gas pipeline leaks; rice and cattle farming	21 times more powerful	+ 250%
Nitrous oxide	3%	jet engines; vehicles; sewage farms; synthetic chemical production	250 times more powerful	+ 16%
Halocarbons	1%	used in industry, solvents and cooling equipment	3000 times more powerful	not natural

Figure 2 *Greenhouse gases – sources and effects*

The enhanced greenhouse effect

The concentrations of several greenhouse gases in the atmosphere (including atmospheric carbon: CO_2 and CH_4) have increased by 25% since 1750.

- Since the 1980s, 75% of CO_2 emissions have come from burning fossil fuels. Most climate researchers believe this is causing increased global temperatures, leading to an **enhanced greenhouse effect**.

- Human activities release natural stores of carbon and nitrogen, which combine with oxygen to form greenhouse gases.
- As global temperatures increase, the level of water vapour in the atmosphere increases; there is greater evaporation and condensation. This causes increased cloud cover, trapping heat in the atmosphere.

Climate patterns – temperature and precipitation

As greenhouse gases naturally help to maintain the Earth's temperature, and also determine the distribution of temperature and precipitation, changing their concentrations is likely to alter these patterns.

See page 61 of the student book for more information on temperature and precipitation patterns.

Regulating the composition of the atmosphere

Photosynthesis is a vital process in regulating the composition of the atmosphere.

- Phytoplankton in oceans sequester CO_2 through photosynthesis, pumping it out of the atmosphere and into the ocean store (see Section 2.3).
- Terrestrial photosynthesis enables plants to sequester CO_2, which is released back into the atmosphere through respiration and decomposition.

Different ecosystems absorb CO_2 due to photosynthesis (Figure 3).

- Anything that affects the level of phytoplankton in the oceans or forested areas will affect the level of carbon sequestration, and the composition of the atmosphere.

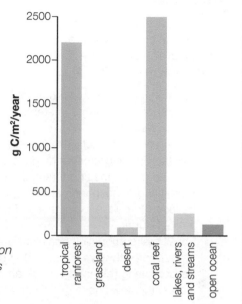

▶ **Figure 3** *Carbon absorption resulting from photosynthesis in different ecosystems*

Soil and carbon

Organic matter is the medium by which carbon passes through the soil system. It supports micro-organisms that maintain the nutrient cycle and break down organic matter, provides pore spaces for infiltration and storage of water, and enhances plant growth (Figure 4). Without carbon, nutrient and water cycles cannot operate properly.

- The amount of organic carbon stored within soil = inputs (plant litter and animal waste) minus outputs (decomposition, erosion and uptake in plant growth).
- The size of the store depends on different biomes.

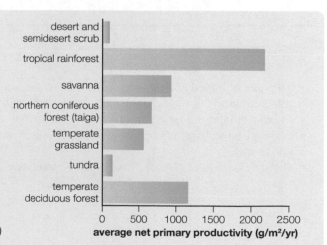

contains many organisms

dark, crumbly, porous

provides air, water and nutrients for organisms and plants

contains carbon and organic matter

improves resilience to wet weather

retains moisture

▶ **Figure 4** *Healthy soils*

Ecosystem productivity

About 1% of solar insolation reaching Earth is captured by photosynthesis and used by plants to produce **biomass** (organic material). The rate at which plants produce biomass is called **primary productivity** (Figure 5).

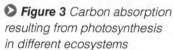

▶ **Figure 5 Net primary productivity** *for different ecosystems (i.e. the amount of biomass produced by plants, minus the energy lost through respiration)*

Fossil fuel combustion

Fossil fuels comprise carbon locked within the remains of organic matter. Most of the world's oil and gas is extracted from rocks that are 70–100 million years old. When burnt to generate energy, stored carbon is released, mainly as CO_2. It has several implications.

Balance

Earth's carbon reservoirs act as sources (adding carbon to the atmosphere) and sinks (removing it). If sources and sinks are equal, the carbon cycle is in **equilibrium** (or **balance**).

- Maintaining a steady amount of CO_2 in the atmosphere helps to stabilise global temperatures.
- However, human activities have increased CO_2 inputs, without corresponding increases in natural sinks (e.g. oceans and forests).
- Increasing atmospheric stores of carbon is widely believed to be the main cause of rising global temperatures.
- Fossil fuel combustion has altered the balance of carbon **pathways** (i.e. flows) and stores: with carbon being released in large amounts from stores, the flows have greatly increased.

Implications for climate

Changes in climate are likely to vary.

- Across Europe, annual average land **temperatures** are projected to increase by more than the global average. The largest increases are expected to be over Eastern and Northern Europe in winter, and over Southern Europe in summer.
- Annual **precipitation** is projected to increase in Northern Europe and to decrease in Southern Europe.
- **Extreme weather events** are likely to increase in intensity and frequency.

Arctic amplification

The Arctic region is warming twice as fast as the global average. This is known as **Arctic amplification**.

- Melting permafrost releases CO_2 and CH_2, increasing their concentration in the atmosphere.
- This leads to increased global temperatures and further melting (i.e. **positive feedback**).

Climate change is altering the **Arctic tundra** because of extensive melting of sea ice in summer, reductions in snow cover and permafrost.

- Shrubs and trees, previously unable to survive, have started to establish themselves.
- In Alaska, the red fox has spread northwards and competes with the Arctic fox for food and territory.

Implications for the hydrological cycle

Projected changes to temperature and precipitation patterns across Europe will impact on the hydrological cycle. In summer, much of Europe's water comes from melting Alpine glaciers. But by 2100, the Eastern Alps and large parts of the Western Alps could be ice-free.

Likely impacts on the hydrological cycle could be:

- Less winter snowfall and rainfall. River discharge patterns could change, with greater flooding in winter and drought in summer.
- As glaciers melt, water flows would result in increased sediment yield.
- Once glaciers have retreated, discharge and sediment yields would fall and water quality decline.

 See page 65 of the student book for information on carbon pathways.

 Ten-second summary

- The concentration of atmospheric carbon influences the natural greenhouse effect and modifies temperature and precipitation patterns.
- Photosynthesis is vital in regulating the composition of the atmosphere.
- Soil carbon storage influences nutrient and water cycles, and is important for soil health.
- Burning fossil fuels affects climate, ecosystems and the hydrological cycle.

 Over to you

Create a mind map to show the impacts of fossil fuel combustion on the balance of the carbon cycle, climate, ecosystems and the hydrological cycle.

You need to know:

- about differences in energy consumption, and about the energy mix
- how access to, and consumption of, energy resources depends on a range of factors
- that energy players have different roles in securing pathways and energy supplies.

Big idea

Energy security is a key goal for countries, with most relying on fossil fuels.

Energy consumption

The amount of energy consumed depends on many things – lifestyle, climate, technology, availability and demand.

- Global consumption continues to rise as countries develop economically.
- There is a close correlation between GDP per capita and energy consumption (Figure 1).

▶ **Figure 1** *The relationship between GDP per capita and energy use, 2011*

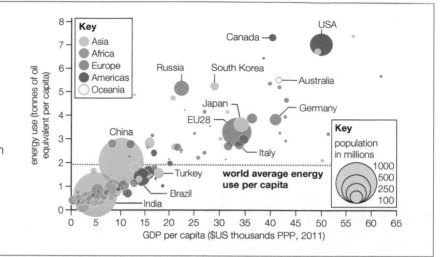

The energy mix

Countries need an **energy mix** of several sources, rather than relying on one.

Primary and secondary energy sources

- **Primary energy sources** are those consumed in their raw form, e.g. burning fossil fuels, nuclear energy (controlling uranium or plutonium to create heat via atomic reaction) and renewable sources (solar, wind, wave energy).
- Primary sources can be used to generate electricity (a **secondary energy source**).

Domestic and overseas sources

- Declining **domestic** North Sea oil and gas reserves have made the UK increasingly dependent on **imported** energy.
- The UK now has an **energy deficit** and is **energy insecure**. Countries with an energy surplus (i.e. exceeding demand) are **energy secure**, e.g. Russia.

Renewable and non-renewable sources

Sources of energy can be classified as:

- **Non-renewable (finite)** e.g. coal, oil, gas. These stocks will eventually run out.
- **Renewable**, e.g. solar, wind, wave power. These are continuous flows of nature, which can be constantly reused.
- **Recyclable**, e.g. reprocessed uranium and plutonium from nuclear power plants and heat recovery systems.

Factors affecting energy consumption

A country's energy mix reflects a range of factors (Figure 2). Figure 2, below and on page 46, compares factors in the UK and Norway.

Factor	UK	Norway
Physical availability	• Dependent on domestic coal until 1970s. • Increased use of North Sea oil and gas after 1970s.	• Hydroelectric power is the natural energy choice. • Much of Norway's oil and natural gas is exported.
Cost	• North Sea reserves became 'secure' alternative to rising price of Middle Eastern oil. • North Sea oil is expensive to extract. Oil and gas stocks are declining, forcing UK to import more.	• HEP supplies 98% of Norway's renewable electricity. • HEP costs are low once capital investment is complete. Transfer of electricity to users is expensive.
Technology	• Current technology and environmental policy make extraction and use of remaining coal unrealistic/expensive. • 'Clean coal' technology exists (i.e. absorbing CO_2 emissions), but has no political support.	• Deepwater drilling technology enabled Norway and the UK to extract North Sea oil and gas.
Political considerations	• Increasing reliance on imported energy affects energy security. • Public concern is growing over fracking and nuclear sites. • Privatisation of UK's energy supply industry means overseas companies (e.g. France's EDF) decide which energy sources will meet UK demand.	• Government prevents foreign companies from owning primary energy source sites (waterfalls, mines, forests). • Some government profit from fossil fuel sales goes to a wealth fund to prepare for a future without fossil fuels and to invest in sustainable projects.

Factor	UK	Norway
Level of economic development	• GDP per capita (PPP) = US$41 200 (2015). • Energy use per capita = 2752 kg oil equivalent (2014).	• GDP per capita (PPP) = US$61 500 (2015). • Energy use per capita = 5854 kg oil equivalent (2014).
Environmental priorities	• UK intends to broaden energy mix with renewable sources and more nuclear power. • In 2015, CO_2 emissions were 7.13 tonnes per capita (down from 11.5 in 1980).	• Norway intends to be carbon neutral by 2050. • In 2015, CO_2 emissions were 11.74 tonnes per capita (up from 11.6 in 1989).
	In 2015, the UK and Norway committed to a 40% reduction in domestic greenhouse gas emissions by 2030, compared to 1990 levels.	

 Figure 2 Factors affecting energy use in the UK and Norway

Changing consumption

Overall energy consumption, as well as fossil fuel consumption, has changed in Norway and the UK (Figure 3).

See page 69 of the student book to find out how the energy mixes of the UK and Norway have changed since 1970/80.

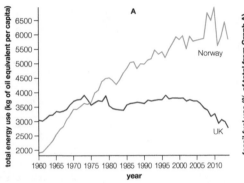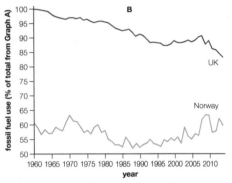

 Figure 3 Overall energy consumption per capita in the UK compared with Norway (**A**) and changing fossil fuel consumption as a percentage of the total (**B**)

Energy players

There are various key players in securing **energy pathways** and supplies (Figure 4). The term 'energy pathway' describes the flow of energy between a producer and a consumer, and how it reaches the consumer (e.g. pipeline, transmission lines, ship).

Energy player	Role	Examples
Energy TNCs (operate across the world)	• Explore, exploit and distribute energy resources. • Own supply lines and invest in distribution, processing raw materials, electricity production and transmission. • Aim to secure profits for shareholders.	• Old players: BP (UK); Shell (UK and Netherlands); Exxon/Mobil (USA). • 'New' players: Petrobras (Brazil); PetroChina Corp (China); Reliance (India); Rosneft, Luckoil, Gazprom (Russia).
OPEC Members are oil producing/exporting countries (e.g. Saudi Arabia). Control 81% of proven world oil reserves.	• Aim to co-ordinate and unify petroleum policies of its members to ensure stabilisation of oil markets.	• From 2012 to 2016, maintaining output at high levels kept oil prices low, possibly to compete with the USA's increased oil production from fracking (which had caused a collapse in global oil prices).
National governments	• Meet international obligations, whilst securing energy supplies for the nation and supporting economic growth. • Regulate the role of private companies and set environmental priorities.	• EDF (France) and China General Nuclear are government-backed energy TNCs involved in developing nuclear power plants in the UK, e.g. Hinkley Point C. • EU governments aim to fulfil CO_2 emissions targets and reduce fossil fuel dependency.
Consumers	• Create demand. Purchasing is often based on price/cost issues, e.g. competitive petrol pricing. • Have some power over oil companies, e.g. by purchasing electric cars, installing solar panels to cut energy costs, etc. • The expansion of nuclear energy, and fracking, are both highly controversial. There have been protests against both.	

⬆ *Figure 4* Key players in securing energy pathways and supplies

Ten-second summary

- There is a close correlation between energy consumption and level of economic development.
- A country's energy mix depends on a range of factors.
- Key players have different roles in securing energy pathways and supplies.

Over to you

Draw a mind map to show how the following can influence a country's energy mix:

- resource availability
- accessibility and affordability
- government policies
- public perceptions.
- technology
- development levels
- environmental concerns

You need to know:

- there is a mismatch between locations of fossil fuel supplies and regions where demand is greatest
- that energy pathways can be disrupted
- that the development of unconventional fossil fuels has costs and benefits.

Big idea

A reliance on fossil fuels to drive economic development is still the global norm.

The location of fossil fuels

Fossil fuels are not distributed evenly; their location is determined by underlying geology. Figure 1 shows the top ten countries in 2014 in terms of oil reserves.

 See page 72 of the student book on how coal, oil and natural gas formed under past geological conditions, which determine where they are found.

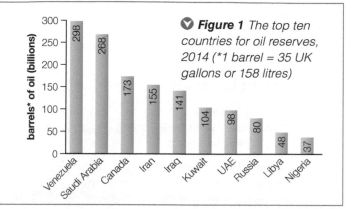

Figure 1 *The top ten countries for oil reserves, 2014 (*1 barrel = 35 UK gallons or 158 litres)*

Fossil fuel demand

As more countries develop, the demand for energy is increasing.

- Fossil fuels still make up 86% of the global energy mix.
- Global energy consumption has increased by 50% since the 1990s. China's rapid economic growth has largely driven this increase (Figure 2).

The problem is the mismatch between where fossil fuels are found and where demand is greatest.

- China's oil reserves are only 10% of the size of Canada's, and China is now the world's second largest oil importer (behind the USA).

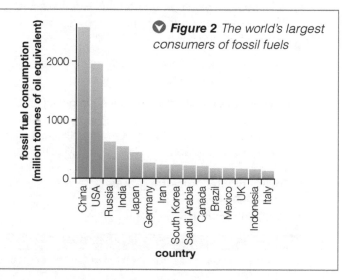

Figure 2 *The world's largest consumers of fossil fuels*

Energy pathways

An **energy pathway** is the flow of energy between a producer and consumer, e.g. pipelines, transmission lines, shipping routes, road and rail.

- Some of the world's largest pipelines carry billions of cubic metres of gas.
- These pathways depend on **multi-lateral** (between many countries) and **bilateral** (between two countries) agreements.
- For security reasons, when companies like Russia's Gazprom export their gas to Europe, they try to avoid **transit states** (through which energy flows on its way from producer to consumer).

Trade flows, shipping routes and disruptions

About half of the world's oil is moved by shipping tankers.

- Much of it goes through eight major **chokepoints** (narrow channels where transport can be disrupted).
- 20% of the world's oil passes through the Strait of Hormuz, a 39 km-wide stretch of water between the Gulf of Oman and the Persian Gulf.
- If chokepoints are blocked, or threatened, energy prices can rise.

Disruption to energy pathways can undermine energy security, resulting from:

- piracy attacks, e.g. along the Strait of Malacca
- attacks on pipelines by militants
- damage to pipelines caused by weather.

Political conflict and energy pathways

The ongoing Syrian conflict has involved:

- Russia and its Shia non-fundamentalist allies
- the USA and its fundamentalist Sunni allies.

Both sides are involved in the battle for control over Syrian territory. Many argue that the key reason for Russia and the USA's involvement is the proposed construction of oil and gas pipelines through Syria into Europe (the world's largest energy market).

Developing unconventional fossil fuels

See page 75 of the student book for more on unconventional fossil fuels.

The world has finite reserves of fossil fuels, which may be reaching an end.

* Canada now exploits tar sands to boost its energy security.
* Shale gas in the USA (produced by fracking) has brought an energy boom.

Unconventional (i.e. new or different) fossil fuels include:

* **deep water oil**
* **tar sands**
* **shale gas**
* **oil shale**.

Canadian tar sands

Canada has the world's largest reserves of tar sands, with three major deposits in Alberta (Figure 3), in ecologically fragile areas.

* Tar sands are extracted by opencast mining.
* Extracted material is crushed and mixed with water, and the bitumen is separated before it can be used.
* Tar sands can also be pumped out. High-pressure steam is injected underground to separate the bitumen from the sand.

Exploiting tar sands involves several key players (below), and brings costs and benefits (Figure 4).

* **Governments** – Alberta regional government and Canada's national government.
* **Oil companies** – Local and international companies include Syncrude/Suncor, Shell, Exxon Mobil and BP.
* **Environmental pressure groups** – e.g. Greenpeace.
* **Local communities**

Figure 3 The location of Canada's tar sands

Costs	Benefits
• 5–10 times more expensive to extract bitumen from tar sands than to produce conventional oil. • Very energy- and water-intensive. • Produces huge quantities of waste (spoil heaps) and toxic wastewater. • Carbon emissions rise due to extraction, production and use. • Carbon absorption falls due to deforestation (removal of the taiga). • Disruption to way of life. • Forest and peat bogs destroyed with a loss of ecosystems and habitats. Reduces resilience of the Taiga. • Decline in caribou populations.	• Offers energy security for Canada and the USA. By 2030, it could meet 16% of North America's oil needs. • Can serve as a fuel stopgap until more renewable and cleaner energy sources become viable. • Environmental protection ensures that mining companies reclaim land disturbed by extraction. • Earns revenues for local and national economies. • Provides new jobs.

Figure 4 The costs and benefits of exploiting tar sands

Ten-second summary

* There is a mismatch between where fossil fuels are found and where demand is highest.
* Energy pathways can be disrupted.
* Unconventional fossil fuels are being developed to boost energy security.
* The development of unconventional fossil fuels has costs and benefits.

Over to you

Use the information in Figure 4 to create a spider diagram showing the:

* social costs and benefits of exploiting tar sands
* implications for the carbon cycle of exploiting tar sands
* impacts on the environment of exploiting tar sands.

You need to know:

- that renewable and recyclable energy could fuel future economic growth but they have costs and benefits
- that there is uncertainty over how 'carbon neutral' biofuels are
- how radical technology could reduce carbon emissions.

Big idea

There are alternatives to fossil fuels but each has costs and benefits.

The UK's changing energy mix

The UK's use of fossil fuels is falling. In order to reduce carbon emissions and create a more secure future (60% of its energy is imported), the UK is committed to decoupling the economy from fossil fuels by:

- increasing renewable energy. According to Friends of the Earth, wind energy could provide 25% of the UK's electricity needs by 2020. Solar power has grown rapidly – by 86% from 2014 to 2015 (Figure 1)
- developing new nuclear power stations, e.g. Hinkley Point C
- reducing energy use through technologies such as LED light bulbs
- recycling energy that would normally be wasted.

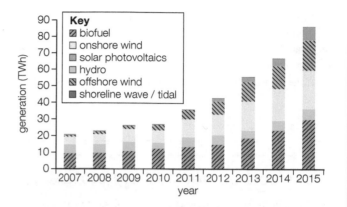

Figure 1 *Growth in the UK's renewable energy generation, 2007–2015. Wind and solar energy have only expanded through government subsidies*

Renewable and recyclable energy sources

In a privatised energy market, energy companies invest when the government guarantees a minimum price per mega-watt hour (MwH), known as the **strike price** (Figure 2).

Figure 2 *Renewable and recyclable energy strike prices per MwH*

Energy type	Strike price per MwH
Renewable	
• Biomass	• £80
• Solar power	• £50–80
• Wind energy	• onshore £80, offshore £115–120
• Wave and tidal energy	• N/A; technology is at research stage
• Hydroelectric power (HEP)	• £100
Recyclable	
• Nuclear power	• Strike price for Hinkley Point C £92.50
• Heat recovery systems, or ground source heat pump	• N/A

Alternatives to fossil fuels – costs and benefits

Nuclear power	Wind power	Solar power
• **Japan** – Before the 2011 earthquake and tsunami, nuclear power provided 27% of Japan's electricity. Earthquake damage to the Fukushima nuclear plant meant Japan closed all its nuclear reactors. Nuclear energy has since been reintroduced to the energy mix. • **The UK – Hinkley Point C** is an £18 billion project, which will provide 25 000 jobs and energy for 60 years. It involves French state-owned EDF and China General Nuclear.	• **Hornsea Project 1** – 190-metre-high wind turbines will provide power for a million homes once completed (2020). Located 121 km off the Yorkshire coast, it will create 2000 construction jobs. • **Quarrendon Fields, Aylesbury** – A wind turbine 25 metres taller than other onshore turbines will supply 2000 homes. Some see it as an eyesore, harmful to birds, with intermittent supply.	• **Chapel Lane Solar Farm, Christchurch** – The UK's largest solar farm, serving 75% of homes in Bournemouth. Cost £50 million and covers an area equivalent to 175 football pitches. Solar and wind energy aren't viable without a high strike price. They consume farmland, which some argue should be producing food.

Figure 3 *Costs and benefits of some alternatives to fossil fuels*

The growth of biofuels

Brazil was the first to produce **biofuel** from sugar cane in the 1970s.

- The bio-ethanol produced was used as a vehicle fuel. It emits 80% less CO_2 than petrol.
- Since 2003, Brazil's use of bio-ethanol has reduced CO_2 emissions by over 350 million tons.

Other countries have followed (Figure 4).

- Malaysia has cleared rainforests to plant oil palms, the EU grows oilseed rape and the USA grows maize – all to produce bio-diesel.
- One downside is deforestation. Another is social unrest, such as in Brazil, where farm workers have lost land to sugar cane and are less able to grow food for themselves. Many move to the cities.

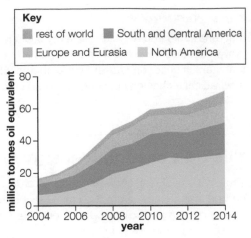

Key
- rest of world
- South and Central America
- Europe and Eurasia
- North America

 Figure 4 Growth in biofuel production, 2004–2014

How 'carbon neutral' are biofuels?

- The plants need pesticide and fertiliser, made from fossil fuels.
- Clearing forest to grow biofuels means the loss of a carbon sink and increased CO_2 emissions from deforestation.
- Biomass needs a fuel to 'kick-start' burning (fossil fuels may be used).

- Most renewable installations (e.g. wind turbines) use energy during their construction, adding to CO_2 emissions.
- The carbon footprint of biofuels varies. Some (e.g. rapeseed) produce more carbon emissions than crude oil.

Reducing carbon emissions: radical technology

Carbon capture and storage (CCS)

This captures CO_2 emissions from coal-fired power stations.

- CO_2 is stored, compressed and transported by pipeline to a well, where it's injected into geological reservoirs, e.g. underground aquifers.
- Theoretically, CCS could cut global CO_2 emissions by up to 19% but it's not currently financially viable.

Hydrogen fuel cells

Once hydrogen is separated from other elements, it provides an alternative to oil.

- Fuel cells convert chemical energy in hydrogen to electricity, with water as a by-product.
- Hydrogen fuel cells are far more energy efficient than petrol engines in vehicles.
- Separating hydrogen from other elements requires energy, which can be provided by renewable sources.

Electric vehicles

Traditionally, problems with electric vehicles include their range (distance travelled before recharging) and price. The range of some models is extending and prices falling. However, batteries can be highly toxic.

 Ten-second summary

- The use of renewable and recyclable energy could help the UK to decouple the economy from fossil fuels.
- Alternatives to fossil fuels have costs and benefits.
- Biofuel production has grown dramatically, but there are questions about how 'carbon neutral' biofuels are.
- Carbon emissions can be reduced by using radical alternatives to fossil fuels but there is uncertainty about these.

Over to you

Close your book. Write down:

- three costs/benefits of renewable/recyclable energy sources
- two issues related to how 'carbon neutral' biofuels are
- one alternative approach to reducing carbon emissions.

You need to know:

- how growing demand for resources has led to changes in land use, affecting carbon stores, the water cycle and soil health
- that ocean acidification is increasing and risks damaging marine ecosystems
- that climate change may increase the frequency of drought.

Big idea

Biological carbon cycles and the water cycle are threatened by human activities.

Deforestation in Madagascar

Growing demands for resources have led to land use changes, which threaten carbon and water cycles. Since the 1950s, Madagascar's tropical forests have been cleared rapidly.

- Growing demand for tropical hardwood, an expanding population and high debt repayments meant that the Madagascan government encouraged farmers to clear more land and grow cash crops to earn foreign currency and help repay its debts.
- By 1985, two-thirds of Madagascar's forest had been lost.

Deforestation has major impacts:

- increasing CO_2 emissions
- reducing terrestrial carbon stores
- affecting the water cycle, soil health, the atmosphere and biosphere.

Impacts on the atmosphere

- Oxygen content reduces. Transpiration rates fall.
- More sunlight reaches the forest floor.
- Reduced evapotranspiration makes it less humid. (Evapotranspiration rates from grasslands are about 30% that of tropical rainforest.)

Impacts on soil health

- Finer particles are washed away. Coarser, heavier material is left behind.
- CO_2 is released from decaying woody material.
- Biomass is lost.
- Soil erosion and increased **leaching** lead to loss of nutrients and minerals.

Impacts on the water cycle

- Infiltration decreases.
- Runoff and erosion increase; rivers carry more sediment.
- Increased discharge leads to flooding with higher flood peaks and shorter lag times.
- Annual rainfall reduces. Seasonality of rainfall increases.

Impacts on the biosphere

- Evaporation from vegetation reduces.
- Less absorption of CO_2 means a reduced carbon store.
- Biodiversity falls.
- Ecosystem services are reduced.
- Biomass is lost.

Converting grasslands to farming

Growth in biofuel crop production (from 2007 to 2015) in the American Midwest reflected growing global demand (see Section 2.7).

- Over 5.5 million hectares of natural grassland (used for cattle ranching) disappeared, matching the rate of deforestation across Brazil, Malaysia and Indonesia.
- The destruction of grasslands has consequences for carbon and water cycles, and soil health (Figure 1).

▶ *Figure 1 The benefits of grasslands and disadvantages of ploughing them for biofuel crops*

Benefits of natural grasslands	Disadvantages of converting grasslands to biofuel crops
Grasslands: • trap moisture and floodwater • maintain healthy soils and absorb toxins • provide cover for dry soils • maintain natural habitats • act as a carbon sink, absorbing CO_2 and releasing O_2 (a 'lung effect') • are a terrestrial carbon store.	• Removing grasslands releases CO_2 from soils. • Ploughing enables soil bacteria to release CO_2. • Biofuel crops need fertiliser and pesticides, producing a net increase in CO_2 emissions. • Biofuel crops need irrigation. • Cultivated soils are liable to erosion, reducing the soil carbon store. • Natural habitats are reduced. • The 'lung effect' is reduced.

Afforestation

Trees provide a vital carbon store, sequestering carbon during photosynthesis.

- Re-planting trees after deforestation, or establishing forests on land not previously forested (**afforestation**), can counter the negative impacts of deforestation.

Ocean acidification

Oceans are a major carbon sink.

- They have absorbed about 30% of all CO_2 produced by human activity since 1800 and 50% of that produced by burning fossil fuels.
- As CO_2 in the ocean increases, pH decreases (i.e. it becomes more acidic). This is known as **ocean acidification** (Figure 2).

▶ *Figure 2 Ocean acidification has lowered the pH of the ocean by about 0.1. It's now 30% more acidic than it was in 1750 (in the early days of the Industrial Revolution)*

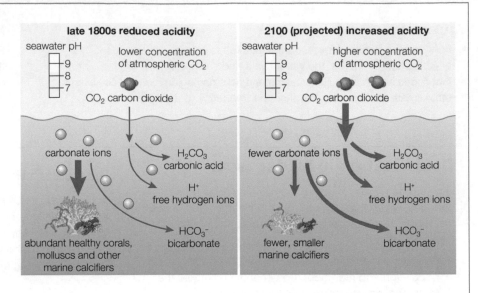

late 1800s reduced acidity

seawater pH

9
8
7

lower concentration of atmospheric CO_2

CO_2 carbon dioxide

carbonate ions

H_2CO_3 carbonic acid

H^+ free hydrogen ions

HCO_3^- bicarbonate

abundant healthy corals, molluscs and other marine calcifiers

2100 (projected) increased acidity

seawater pH

9
8
7

higher concentration of atmospheric CO_2

CO_2 carbon dioxide

fewer carbonate ions

H_2CO_3 carbonic acid

H^+ free hydrogen ions

HCO_3^- bicarbonate

fewer, smaller marine calcifiers

The problems for coral

See page 86 of the student book for information on ecosystem services.

Coral polyps get their colour from algae that live in their tissues.

- Algae provide the coral's food through carbohydrates produced during photosynthesis.
- Coral polyps live within a narrow temperature range – ideally 23–29°C. If water becomes too warm, algae are ejected and coral turns white, called **coral bleaching**.
- If CO_2 emissions continue at their current rate, the pH of ocean surfaces could fall to 7.8 by 2100, dissolving coral skeletons and causing reefs to disintegrate.
- Increased acidification risks crossing the **critical threshold** for the health of coral reefs and other marine ecosystems, and vital ecosystem services they provide will be lost.

Increasing drought

One of the likely impacts of climate change is more frequent floods, droughts and heat waves.

- Changes in ocean currents and atmospheric circulation could affect patterns of precipitation, evapotranspiration and temperature, as climate belts move.
- According to climatologists, climate belts are changing faster, with increasing warming.
- Certain regions (e.g. mid- and high-latitude regions) will undergo more change than tropical and sub-tropical regions.
- The coldest zones of the planet are decreasing in extent and dry areas are increasing.

Drought in the Amazon

The Amazon Basin suffered severe droughts in 2005, 2010, and 2014/15. The Basin plays a key role in the Earth's carbon cycle, holding 17% of the terrestrial vegetation carbon store. A study of the 2010 drought showed that:

- trees died and growth rates declined
- it shut down the Amazon's function as a carbon sink
- forest fires burnt trees and litter, releasing CO_2.

As climate change increases temperatures and alters rainfall patterns across South America, the Amazon rainforest could change from a carbon sink to a carbon source, accelerating global warming.

Ten-second summary

- Increasing demand for resources threatens the carbon and water cycles.
- Increased ocean acidification risks crossing the critical threshold for healthy marine ecosystems.
- Drought impacts on the health of forests as carbon stores.

Over to you

List ways in which land use changes in Madagascar and the American Midwest are making the impacts of climate change worse on:

a the hydrological cycle
b the carbon cycle
c the health of soils.

You need to know:

- that forest loss has implications for human well-being
- how increasing temperatures affect the water cycle
- how threats to ocean health threaten human well-being.

Big idea

There are implications for human well-being from the degradation of the water and carbon cycles.

Forest loss and impacts on well-being

Huge areas of rainforest in South-East Asia, Latin America and Africa have been destroyed to create oil palm plantations in order to produce palm oil, the most commonly produced vegetable oil. In the process:

- Vast amounts of carbon have been released into the atmosphere.
- Local communities and animals have suffered due to the deforestation.
- Many smallholders and indigenous people (who depend on rainforests for everything they need) are driven away by palm oil producers.
- The loss of biodiversity and habitat endangers many species, e.g. the orang-utan.

Protecting forest stores?

In 2011, Indonesia's President declared a 'forest moratorium' to reduce deforestation.

- Permits were no longer issued for clearing primary forest or peatland for timber, pulp or palm oil.
- By 2013, emissions had fallen by 1–2.5%.
- In 2015, the moratorium was extended to help Indonesia reduce its emissions by 26% by 2020.

The moretorium's effectiveness is limited.

- Clearance permits issued before 2011 went ahead.
- Illegal logging remains a problem.
- It only reduced clearance by 15%.

Forest recovery

Globally, forested areas fell by 3% between 1990 and 2015. This may be changing.

- From 2010 to 2015, an average of 7.6 million hectares of forest were lost every year, but 4.3 million hectares were gained (a net annual loss of 3.3 million hectares – half that of the 1990s).
- Temperate forest areas (covering many HICs) have increased, but tropical forests (covering many LICs) have decreased.
- China aimed to increase its forested area by 23% from 2015 to 2020.
- Brazil has halved its rate of deforestation since 2000.

 See page 89 of the student book on Kuznet's Curve (a concept where environmental concerns move up the political agenda) and Attitudes and actions (how consumers respond to environmental issues such as palm oil production).

Climate change and Yukon

Part of Yukon (Figure 1), in north-western Canada, lies within the Arctic Circle. Temperatures there have risen sharply, which has a range of implications (Figure 2).

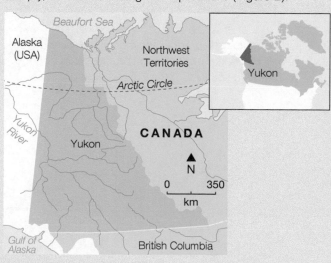

Figure 1 *The Yukon*

Evaporation and atmospheric water vapour	will increase.
Precipitation	from 1950 to 1998, more fell as rain in spring and less as snow than previously. Scientists say annual precipitation will increase by 5–20% by 2100.
Snowmelt	now begins earlier.
Snow cover	is decreasing.
River regimes	peak flows are earlier (due to earlier snow melt).
Total ice area	shrank by 22% from 1958 to 2008.
Streamflow	is decreasing as glaciers recede.
Permafrost	is thawing, so water penetrates deeper into the soil, increasing the amount of groundwater.
Inflows to the Yukon River	have increased by 39% since 2000.

Figure 2 *Implications of climate change in Yukon*

Changing precipitation patterns

See page 90 of the student book on the uncertainty of climate projections.

Climate models predict that as temperatures rise, precipitation patterns will change.

- Existing patterns will strengthen ('wet gets wetter, dry gets drier'). Warmer air traps more water vapour. Climatologists expect more water to fall in regions that are already wet. Because the Earth is a closed system, an increase in dry areas is also expected.
- As atmospheric circulation changes, storm tracks will shift further towards the Poles.

Ocean health – threats and impacts

Loss of mangroves

Mangrove forests provide many benefits (Figure 3), e.g.

- stabilising coastlines against erosion
- providing protection against extreme weather (e.g. storm winds, floods, tsunami)
- providing fish nurseries.

Globally, half of all mangrove forests have been lost since 1950. Clearing them for tourism or aquaculture has accounted for over 25% of this.

Loss of food

520 million people depend on fisheries for food and income. Climate change is altering the distribution and productivity of species, food webs and biological processes.

 Figure 3 *Mangroves protect coastlines and coastal communities*

- Warming waters in the North Atlantic are killing cold-water plankton (food for cod).
- Arctic krill (food for whales) are declining by up to 75% per decade in parts of the Southern Ocean.
- Ocean acidification and warming oceans are leading to coral bleaching, affecting food sources and incomes for people living in coastal communities.

Loss of tourism

Damage to coral reefs can directly impact on the income that local people derive from tourism. The main cause of damage is climate change, plus pollution and oil spills.

 Ten-second summary

- Deforestation impacts on human well-being.
- Some forest stores are being protected and are expanding in some areas.
- Climate change has impacts on precipitation patterns, river regimes and water stores.
- Degradation of ocean health threatens the well-being of many coastal communities.

Over to you

Draw a spider diagram to show how threats to ocean health could affect coastal communities that depend on marine resources.

You need to know:

- that future emissions, atmospheric concentration levels and climate warming are uncertain
- that adaptation strategies are ways to cope with climate change
- that mitigation strategies could help to rebalance the carbon cycle.

Big idea

Further planetary warming risks a large-scale release of stored carbon, requiring responses from different players at different scales.

Future emissions, atmospheric concentration levels and climate warming

As energy consumption rises, greenhouse gas emissions are expected to continue to increase.

- It is uncertain by how much they will increase or where the greatest increases will be.
- Climate scientists use a range of scenarios to show projected atmospheric greenhouse gas concentrations (Figure 1).

Climate models predict that surface temperatures will continue to rise 2–6°C between 2000 and 2100.

- Temperatures are expected to warm more rapidly over the Northern Hemisphere (with more land than ocean).
- Some regions (e.g. the Arctic) will see larger increases than the global average.

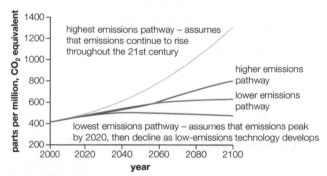

Figure 1 *Projected scenarios for atmospheric greenhouse gas concentrations*

Why is future climate change so uncertain?

Physical factors

- **Oceans** and forests act as **carbon sinks** and store heat. The oceans' response to higher levels of greenhouse gases and temperatures will affect climate for hundreds of years.
- Globally, the area of forested land is falling, due to human action.

Human factors

- **Economic growth** – the rate of growth of emissions hasn't followed global economic recovery. The rate fell to 0.5% by 2014, but this still meant that total emissions reached a new record.
- **Energy sources** – renewable sources accounted for 66% of increased electricity production in 2015.
- **Population change** – Increasing affluence in emerging economies means more emissions.

Feedback mechanisms

Feedback can either dampen (**negative feedback**) or amplify (**positive feedback**) responses to external factors that affect global climate. Both mechanisms below increase the concentration of atmospheric greenhouse gases.

Carbon release from permafrost

Melting permafrost releases carbon as CO_2 and methane.

Carbon release from peatlands

Peat stores large amounts of carbon.

- Warming causes peat to dry out and increases the rate of decomposition.
- Carbon is emitted as methane.

Tipping points

A climate tipping point is a critical threshold: a small change in the global climate system can transform it into a different state. Tipping points include:

- **Forest die back** – rainfall in the Amazon Basin is largely recycled from moisture within the forest. In droughts, trees die back. If die back stops the recycling of moisture (the tipping point), further die back will result.
- **Changes to the thermohaline circulation** – melting northern ice sheets could block and slow the 'conveyor belt' of warm water from the Tropics (see Section 1.2 on thermohaline circulation). This could create a tipping point in ocean circulation, which *might* affect global temperatures.

Facing the future

There are two approaches (Figures 2 and 3):

- **Adaptation strategies** – ways to live with the impacts of climate change.
- **Mitigation strategies** – which re-balance the carbon cycle and reduce the impacts of climate change.

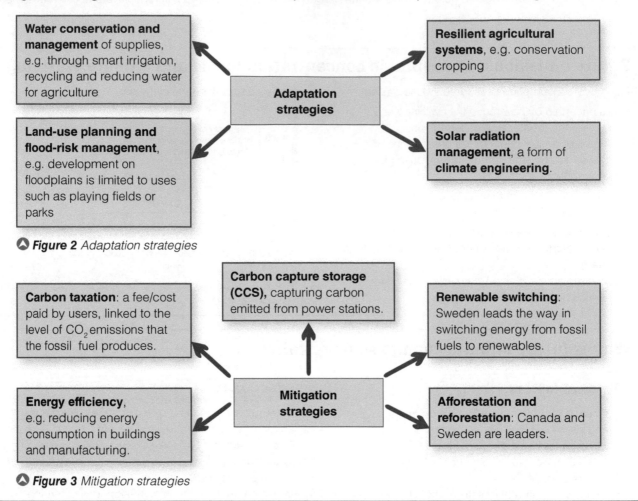

Water conservation and **management** of supplies, e.g. through smart irrigation, recycling and reducing water for agriculture

Land-use planning and flood-risk management, e.g. development on floodplains is limited to uses such as playing fields or parks

Adaptation strategies

Resilient agricultural systems, e.g. conservation cropping

Solar radiation management, a form of **climate engineering**.

⬆ **Figure 2** *Adaptation strategies*

Carbon taxation: a fee/cost paid by users, linked to the level of CO_2 emissions that the fossil fuel produces.

Carbon capture storage (CCS), capturing carbon emitted from power stations.

Renewable switching: Sweden leads the way in switching energy from fossil fuels to renewables.

Energy efficiency, e.g. reducing energy consumption in buildings and manufacturing.

Mitigation strategies

Afforestation and reforestation: Canada and Sweden are leaders.

⬆ **Figure 3** *Mitigation strategies*

Global agreements and national actions

The most significant climate agreement so far is the **2015 Paris Agreement**. 195 countries adopted this legally binding global climate deal. It set out to:

- limit the average global temperature increase to 1.5°C above pre-industrial levels
- strengthen adaptation and resilience in dealing with the impacts of climate change
- support developing countries in reducing emissions and adaptation.

Actions and attitudes

Achieving agreement about actions on climate change depends on the co-operation of many groups.

- **Governments** may disagree on the best ways to progress. Some fear that curbing emissions universally might hamper economic growth for developing countries.
- Reduced emissions could increase manufacturing costs for **TNCs**.
- **People** at risk from rising sea levels view climate change as more urgent than others.

 Ten-second summary

- Future climate change is uncertain due to physical and human factors, feedback mechanisms and tipping points.
- Adaptation and mitigation strategies offer different approaches to the future.
- Global agreements are needed to cope with further planetary warming.
- Actions and attitudes towards climate change vary.

 Over to you

Read pages 95 and 96 in the student book and complete a table showing the costs and benefits of:

a adaptation strategies
b mitigation strategies.

Chapter 3
Superpowers

What do you have to know?

This chapter studies how superpowers have developed and how their pattern of dominance has changed over time. They have a very significant impact on the global economy, global politics and the environment, while their spheres of influence are frequently contested.

The specification is framed around three enquiry questions:

1 What are superpowers and how have they changed over time?
2 What are the impacts of superpowers on the global economy, political systems and the physical environment?
3 What spheres of influence are contested by superpowers and what are the implications of this?

The table below should help you.

- Get to know the key ideas. They are important because 12-mark questions will be based on these.
- Copy the table and complete the key words and phrases by looking at Topic 7 in the specification. Section 7.1 has been done for you.

Key idea	Key words and phrases you need to know
7.1 Geopolitical power stems from a range of human and physical characteristics of superpowers.	superpowers, emerging powers, 'hard' to 'soft' power spectrum, Mackinder's geo-strategic location theory
7.2 Patterns of power change over time and can be uni-, bi- or multi-polar.	
7.3 Emerging powers vary in their influence on people and the physical environment, which can change rapidly over time.	
7.4 Superpowers have a significant influence over the global economic system.	
7.5 Superpowers and emerging nations play a key role in international decision-making concerning people and the physical environment.	
7.6 Global concerns about the physical environment are disproportionately influenced by superpower actions.	
7.7 Global influence is contested in a number of different economic, environmental and political spheres.	
7.8 Developing nations have changing relationships with superpowers with consequences for people and the physical environment.	
7.9 Existing superpowers face ongoing economic restructuring, which challenges their power.	

You need to know:

- how geopolitical power can be assessed in both 'hard' and 'soft' ways.

Big idea

Geopolitical power can be measured in 'hard' and 'soft' ways.

Hard and soft power

'Soft power' can be defined as power that arises from three of a country's resources.

- **Culture** – a country has soft power if its culture (e.g. BBC programmes such as *Top Gear*) is attractive to others.
- **Political values** – e.g. its democracy and overseas image.
- **Foreign policies** – where these have moral authority and support.

Soft power contrasts with the idea of '**hard power**', which shows, at its extreme, a country's will or influence through military force.

Because there are so many ways of displaying power, it's best to think of hard and soft power as a spectrum (Figure 1).

Figure 1 *The hard- and soft-power spectrum, showing the range of actions that countries can use to show their power*

Hard power				Soft power
Military force or its threat	Economic sanctions and diplomatic actions	Coercive policy, e.g. tied aid or trade agreements	Political influence, moral authority, economic influence	Cultural attractiveness

Creating a soft-power index

In 2012, the UK-based magazine *Monocle* used a range of data to produce a soft-power index (Figure 2).

- Data included overseas aid contributions, income inequality, democracy and personal freedoms.
- Combined, the ranking showed that the UK topped the index in 2012 (the year of the London Olympics).
- Countries where personal freedoms were lacking (e.g. China, Russia and India) did less well.

Figure 2 *The top ten countries in the* Monocle *2012 'Soft power' rankings. All are democracies, have well-known cultural exports and are regarded as 'fair' in their foreign policies*

Rank	Country	Score
1	UK	7.29
2	USA	6.99
3	Germany	6.48
4	France	6.47
5	Sweden	5.75
6	Japan	5.61
7	Denmark	5.60
8	Switzerland	5.55
9	Australia	5.53
10	Canada	5.42

How effective is soft power?

Many countries rely on soft power for overseas influence because it is attractive and effective. A key example of this is using an event to boost global brand, e.g. hosting the 2012 Olympic Games in the UK.

Other countries do add to their hard power, though this is expensive in money and lives. China has recently expanded its military power and range of weapons to increase its global influence, adding to its economic power.

Ten-second summary

- Power can be seen on a continuum from soft power to hard power.
- Hard power can involve the use of military force.
- Soft power arises from a country's culture, political values and foreign policies.

Over to you

Study Figure 1 and draw a spider diagram to show how the UK has exerted its power in the years since 2000.

You need to know:

- how superpowers are defined
- how human and physical characteristics affect superpower status
- about Mackinder's geo-strategic location theory.

Big idea

Geopolitical power results from a range of human and physical characteristics.

What does 'superpower' mean?

The word 'superpower' was first used at the end of the Second World War to describe global **influence** held by the British Empire, the USA and the USSR (the former Soviet Union).

- Since then, the world has changed significantly and only the USA remains a major **power**. It has had so much power in recent decades that it has become a **hegemon** (supreme power).
- In the 21st century, the BRIC nations (Brazil, Russia, India and China) are all aiming for superpower status.

Seven factors make up superpower status.

1 Physical size and geographical position

The larger the country, the more resources and the larger influence it tends to have.

- For example, Russia is the world's largest country, controlling significant resources.
- Alongside Canada (the second largest country), Russia has huge influence in the Arctic region.

2 Economic power and influence

The world's largest economies (Figure 1) gain influence based on their:

- Ability to **control investment** – most FDI is targeted at these countries because TNCs are attracted by a likely higher profit.
- **Powerful currencies** – the US dollar and the euro are seen as relatively 'safe' for investment alongside the UK pound, the Swiss franc and the Singapore dollar.
- Ability to determine global **economic policies** – e.g. through membership of IGOs (i.e. the G20) or trading blocs (i.e. the EU).

▶ **Figure 1** *The world's ten largest economies in 2015; the percentage of global GDP produced is a factor influencing superpower status*

Rank	Country	Percentage of global GDP produced
1	USA	22.4%
2	China	13.4%
3	Japan	6.2%
4	Germany	4.9%
5	UK	3.9%
6	France	3.7%
7	Brazil	2.9%
8	Italy	2.8%
9	Russia	2.64%
10	India	2.63%

3 Demographic factors

Population size (Figure 2) can be a key to economic success and thus global influence.

- China has used its large labour force to generate economic growth, and its large market to attract TNCs.
- There are exceptions, e.g. Singapore's population is around half of London's yet it has a major influence on South Asia's economy.
- China's and India's large populations are part of their recent success and are likely to remain so.

▶ **Figure 2** *The world's ten most populous countries in 2015*

Rank	Country	Population
1	China	1.36 billion
2	India	1.25 billion
3	USA	322 million
4	Indonesia	256 million
5	Brazil	204 million
6	Pakistan	199 million
7	Nigeria	182 million
8	Bangladesh	169 million
9	Russia	146 million
10	Japan	127 million

4 Political factors

In a globalised world, few countries hold much global influence on their own and membership of IGOs such as the **OECD** is key to gaining global influence.

- Until recently, the **G8** was the most influential group.

- However, the global shift and growth of Asian economies makes the **G20** more significant now.
- G20 membership includes all G8 members, plus the BRICs and the EU – in total, representing half the world's population.

5 Military strength

Military strength has historically been important for gaining (and maintaining) superpower status and is crucial for displaying hard power.

- The USA is still the leader in military spending and **global reach** of its weapons. US military spending is vast relative to that of others (Figure 3).
- Membership of the UN Security Council can be seen as the ultimate status in military power. Its five permanent members (the USA, UK, Russia, China and France) work to approve military intervention.

However, in the 21st century, military size could become less significant in global influence as defence budgets are widely cut.

1st
USA
US$597.5 billion

2nd
China
US$145.8 billion

3rd
Saudi Arabia
US$81.9 billion

 Figure 3 *The world's three largest military budgets in 2016*

6 Cultural influence

Increased globalisation and the resultant cultural diffusion spread via the media and global TNCs, which represent a key soft power element of superpower status.

- US multimedia TNCs such as Disney, News Corporation and Time-Warner all dominate global culture in spreading their films, print and music.
- Their influence also gives them political strength, e.g. in the way News Corp seeks to support certain political parties.

Cultural diffusion is the spread of one culture to another as a result of globalisation. See page 177 of the Year 1 student book.

7 Physical resources

Having resources does not necessarily guarantee economic development but for some superpowers access to resources such as oil or metals has certainly helped. It provides leverage over others, e.g. the influence of OPEC countries in setting global oil prices.

The changing centre of power

The late 20th and early 21st centuries have seen shifts in the balance of power and **geopolitics** (the influence of geography on politics, particularly international relations).

- Mackinder's **heartland** (or **geo-strategic location**) theory argued in 1904 that the countries that controlled Europe and Asia (the world's biggest landmass and 'heartland') would control the world (Figure 4).
- Mackinder believed that the heartland could shift, depending on different factors, especially as a result of sea power (a type of military strength).
- The heartland's position has changed and we could see a return to the dominance of Asia in the near future.

 Figure 4 *Mackinder's heartland. The further away from the heartland a country is, the less influence it would have*

See Figure 9 on page 107 of the student book for a map showing the changing centre of gravity of power since AD1.

 Ten-second summary

- A superpower is a country projecting significant global power and influence.
- Seven human and physical factors contribute to superpower status.
- There has been a global shift in power and Mackinder's 'heartland' geo-strategic location theory helps to illustrate this.

 Over to you

Draw a mind map to show the significance of each of the seven factors in shaping superpower status.

You need to know:

- that superpower influence can change over time
- that during the 19th and early 20th centuries, direct colonial rule maintained power
- that in the late 20th and early 21st century, indirect control and neo-colonial power has become more important.

Big idea

Patterns of power change over time and can be uni-polar, bi-polar or multi-polar.

The multi-polar world

The 19th century was a **multi-polar** world with no single superpower, although the British Empire was the largest. Empires were formed through **colonialism**, where an external nation takes direct control of a territory, sometimes by force.

The emergence of a bi-polar world

British colonialism gave the UK political and economic power until rapid industrialisation of the USA and the creation of the USSR challenged its dominance.

- By 1945, a **bi-polar** (two-sided) world had emerged, with the USA and USSR established as global superpowers.
- The period of balanced power (1945–1991) between the two resulted in the 'Cold War'.

The USA and USSR maintained their influence militarily, politically, culturally and economically.

Military influence

Military power was a significant factor in this bi-polar world. As the balance of power shifted between the two superpowers, new alliances were formed.

- The Warsaw Pact was a military pact formed by countries supporting the USSR. The Council for Mutual Economic Assistance was formed for economic strength.
- NATO was formed of countries supporting US and Western European influences.

Political influence

Political influence became more important after 1945, although the threat of nuclear weapons also prevented open conflict from breaking out.

- The 'Iron Curtain' (Figure 1) was the heavily defended border between Eastern and Western Europe. (At the time, it formed a dividing line between capitalist and communist countries).
- Moscow held significant influence across Eastern Europe from economic planning to military operations.
- NATO used its influence to help maintain peace during the Cold War era.

Cultural influence

The Cold War was based on propaganda rather than military conflict.

- In the USA, the McCarthy Trials were designed to expose anybody with communist leanings, with daily reports on TV and radio.
- Certain films also generated suspicion of communists.

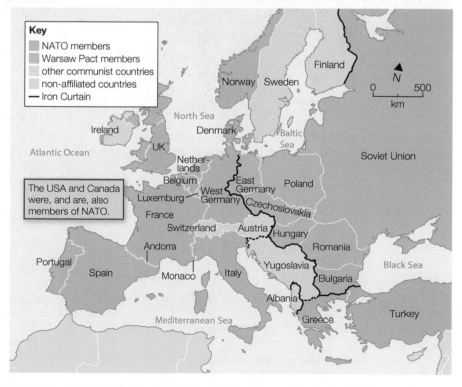

> **Figure 1** The extent of NATO across Europe and the position of the Iron Curtain during the Cold War

Economic influence

After 1945, the USA adopted its Marshal Plan, an aid programme to extend its economic influence and help strengthen war-torn Western European countries (Figure 2).

- Aid was used to rebuild war damage, promote economic development and prevent poverty, which was believed to underpin communist influence.
- Later, the USA provided inward investment to the 'Asian Tigers' (e.g. Singapore) to enable economic growth and prevent further spread of communism.
- Known as **neo-colonialism** ('new' colonialism), this influence has greatly accelerated global development. How much was invested in economic development is disputed and a large proportion was actually spent on military hardware (Figure 3).

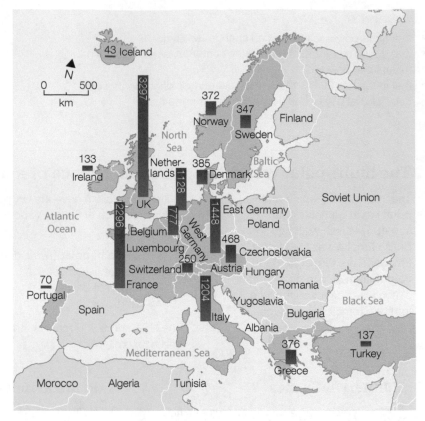

Figure 2 US aid (in US$ millions) given to European countries as part of the Marshal Plan, 1948–1952

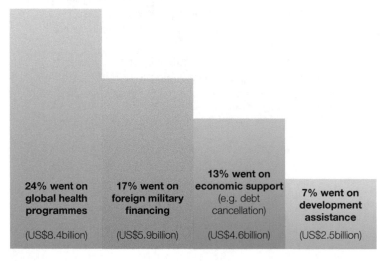

| 24% went on global health programmes | 17% went on foreign military financing | 13% went on economic support (e.g. debt cancellation) | 7% went on development assistance |
| (US$8.4billion) | (US$5.9billion) | (US$4.6billion) | (US$2.5billion) |

Figure 3 The distribution of US aid in 2014 granted for development, although a large proportion was spent on military strength

The rise of China

Different patterns of power bring varying degrees of geopolitical stability and risk. The fall of the Berlin Wall (1989), the end of East Germany's communist government and the collapse of the USSR (1991) left the US leading a **uni-polar** world.

- Since then the US economic, cultural, political and military strength has been virtually unrivalled.
- Recently, China has sought greater influence, challenging the USA.
- China could overtake the USA but it faces domestic problems linked to its lack of democracy and censorship of the population.

Ten-second summary

- The 19th century was a multi-polar world resulting from colonial rule.
- By 1945, a bi-polar world left the USA and USSR as global powers, maintaining influence militarily, politically, economically and culturally.
- By the early 1990s, the USA was the sole superpower in a uni-polar world.
- This is now being challenged by China's rapid economic growth.

Over to you

1 Draw four large overlapping circles (a Venn diagram) to show how a superpower(s) can maintain power politically, militarily, culturally and economically.
2 Which of the four circles do you think is **a** most important and **b** least important, and why.

You need to know:

- that emerging countries are increasingly important to global economic and political systems
- that each of the BRICs has strengths and weaknesses
- how development theory can help explain changing patterns of power.

Big idea

Emerging powers vary in their influence on people and the physical environment.

Rising economic superpowers

The 21st century has brought enormous political and economic changes.

- China, India and Brazil are emerging as major economies.
- Russia is re-emerging as an economic and political power.
- The EU has expanded to become the world's largest trading bloc, competing with the USA for global economic dominance.
- Emerging economies in Asia are likely to be major global players by 2050.

Global governance

There have been changes in the dominating powers of global governance.

- New players have emerged to promote reduced greenhouse gas emissions at the annual UN Climate Change Conference.
- The 2015 meeting saw high numbers of delegates from China, France, Canada and Russia, showing their commitment to emissions reductions.

New BRICs on the block

By 2050, it is likely that a new era of superpowers will begin, as the **BRICs** (Brazil, Russia, India, China, along with South Africa) challenge the uni-polar world dominated by the USA (Figure 1).

Assessing the BRICs

Each of the BRICs has evolving strengths and weaknesses that could inhibit or advance its economic and geopolitical role in the future (Figure 2).

	USA	Brazil	Russia	India	China	South Africa
HDI	0.916	0.755	0.798	0.609	0.727	0.407
GDP per capita US$ PPP	55 837	15 359	24 451	6 089	14 239	13 165
Internet users (% pop)	88.2	60.1	71.3	34.8	52.2	48.9
GDP from agriculture (%)	1.6	5.8	4.0	17.9	9.7	2.5
Population growth rate (%)	0.77	0.8	0.19	1.25	0.44	−0.48

Figure 1 *Key indicators in 2014 for the five BRICs countries, in comparison with the USA*

For information about China and India, see pages 114–115 of the student book.

	Brazil	Russia
Economic	• Important regionally; produces half of South America's GDP. • Relies on primary products for export.	• The 9th largest global economy but high dependency on oil and gas makes it vulnerable to price fluctuations. • High inequality: 35% of wealth was owned by 110 people in 2014 • The poorest 20% share 3% of Russia's GDP.
Political	• Less stable in recent years, with accusations of corruption and protests since 2013.	• Reduced global influence since 1991, though its role in Syria (2012–2017) has rebuilt this. • Maintains political influence over former USSR republics.
Military	• Accounts for over 60% of South America's total military budget but is the least significant of the BRICS.	• Military spending has increased yet naval and aircraft stock is ageing.
Cultural	• Global reputation for football. • Famous for the Rio Carnival.	• Russian is spoken little beyond the borders of the former USSR. • A large cultural tourist industry.
Demographic	• Has half of South America's population. • Has a young population, but it is beginning to age. • Fertility rate of 1.8.	• Population declined during 1991–2015. • Now a tiny natural increase with low fertility rate (1.78).
Environmental	• High biodiversity with 13% of all global species. • Supports global initiatives (e.g. UN Conference on Climate Change) and is a world leader in biofuels. • Deforestation, illegal poaching and pollution are major problems.	• Pollution remains from industrialisation due to deforestation, mining and toxic waste spillages.

Figure 2 *Brazil and Russia have strengths and weaknesses as superpowers that will determine the extent of their global influence in the future*

Explaining changing patterns of power

Three theories help to explain how some countries rise and fall in power.

Wallerstein's world systems theory

Wallerstein developed a theory in 1974 to help explain capitalist world systems and the development gap (Figure 3). He based this on the following points.

- The world's economic **core regions** (HICs) drive the world economy.
- They import, process, add value to and profit from processing primary materials from **peripheral regions** (developing countries).
- This largely stems from unequal trading between former colonial rulers and colonies.
- Unequal trade patterns persist today. Exports of primary products dominate the economies of developing countries whilst core regions dominate ownership of production lines, dictating what is produced and by whom.
- Currently, the Western 'core' owns and consumes 75% of global goods and services.

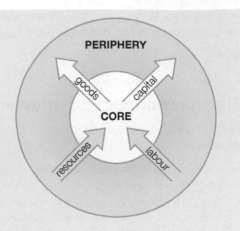

△ **Figure 3** *Wallerstein's world systems theory shows the interactions of capital, goods, resources and labour between the core and periphery areas*

Modernisation theory

In the 1940s, the USA viewed advancing communism as a major threat and promoted modernisation theory. The idea was that to deliver capitalism, modern institutional reform was needed and that capitalism is the solution to poverty.

- The establishment of the IMF and World Bank helped to achieve reform, focusing on currency stability and development loans (see Section 3.5).
- Without reform, the poverty trap would remain and developing economies would be held back by traditional family values.
- In response to advancing communism, US investment was targeted on countries bordering China and the USSR, e.g. Japan, India, South Korea.

Dependency theory

This theory argues that the **dependency** of developing countries on wealthier nations is the cause of poverty.

- Trade patterns involve the export of primary resources to developed nations in return for manufactured goods.
- Tariffs are added to processed imports, so the **terms of trade** are unfavourable to developing countries.
- Developing countries can't process or add value to primary goods so profits are low. This deters investment and maintains poverty. These countries become trapped in a vicious cycle (Figure 4).

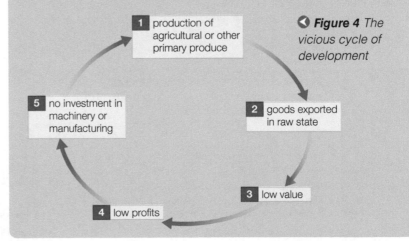

◁ **Figure 4** *The vicious cycle of development*

1 production of agricultural or other primary produce

2 goods exported in raw state

3 low value

4 low profits

5 no investment in machinery or manufacturing

 Ten-second summary

- A number of emerging countries (and the EU) are increasingly important in political and economic terms.
- However, each has strengths and weaknesses that may hinder their rise to superpower status.
- Changing patterns of power can be explained by three different development theories.

Over to you

Look at Figure 2 (on page 63) and refer to pages 114–115 in the student book. Rank the characteristics in order of significance to decide which BRIC is likely to become a superpower.

You need to know:

- how superpowers influence the global economy through IGOs promoting free trade and capitalism
- that TNCs are major players in the global economy with economic, technological and cultural influence.

Big idea

Superpowers have significant influence over the global economic system.

The influence of IGOs

The World Bank, IMF, WTO and WEF all promote free trade and capitalism.

- Fear of communist expansion dominated US foreign policy after the Second World War. As a result, modernisation theory influenced global development, promoting capitalism.
- The IMF and World Bank were set up to help achieve this.

The World Economic Forum (WEF)

A Swiss not-for-profit organisation founded in 1971, the WEF promotes public-private co-operation at its annual forum in Davos.

- Its aims are to bring together businesses and wider society to improve the world, discussing wider issues, from corruption and terrorism to economic systems and social issues.
- Its members think internationally, encouraging governments to promote global links.

The International Monetary Fund (IMF)

Founded in 1944, the IMF:

- aims to stabilise global currencies
- provides loans to help developing countries reduce poverty (and prevent communism)
- sets up Structural Attachment Programs (SAPs) as a condition of loans in order to promote capitalism within the country.

Eight countries control 47% of the votes between them.

The World Bank

Founded in 1944, the World Bank:

- aims to support capitalism by financing project loans to developing countries
- aims to eliminate poverty whilst implementing sustainable goals
- provides finance following natural disasters and humanitarian emergencies.

The USA controlled 16.5% of the votes in 2016.

The World Trade Organisation (WTO)

The WTO focuses on trade and its rules, which ensure that capitalism thrives.

- It aims to free up global trade and reduce trade barriers by negotiating free-trade agreements.
- Its current work includes poverty reduction programmes, such as removing farm subsidies in developing countries to stimulate efficient production. In fact, cheaper imports then undercut local farmers, forcing them out of business.
- A combination of globalisation and WTO agreements has led to an explosion in global trade since 1950.

TNCs as global players

TNCs are not recent features (the East India Company ran most of India in the 18th and 19th centuries). But the growth of TNCs as global players *is* recent. However, changes are taking place.

- Top TNCs are becoming more international (Figure 1). In 2006, six of the largest ten TNCs were American. By 2015, this figure had fallen to three. This reflects the growing influence of Chinese TNCs.
- Most TNCs are publicly owned corporations, driven by profit. However, Chinese TNCs are state-led and operate commercially, while returning all profits to the state.

2006 rank	2015 rank	Company name	Revenue (US$ billions)	Country of origin	Nearest GDP
2	1	Walmart Stores	485.65	USA	Sweden
-	2	Sinopec	433.31	China	Belgium
3	3	Royal Dutch Shell	385.63	UK/Netherlands	Norway
-	4	PetroChina	367.85	China	UAE
1	5	Exxon Mobil	364.76	USA	UAE
4	6	BP	334.61	USA	Egypt
8	7	Toyota Motor	248.95	Japan	Chile
-	8	Volkswagen	244.81	Germany	Chile
-	9	Glencore	209.22	Anglo-Swiss	Portugal
-	10	Total	194.16	France	Vietnam

Figure 1 *The top ten TNCs by revenue in 2015*

Whilst the top TNCs are becoming more international, a small number of countries still dominate (Figure 2) and increasingly Asian TNCs are gaining influence.

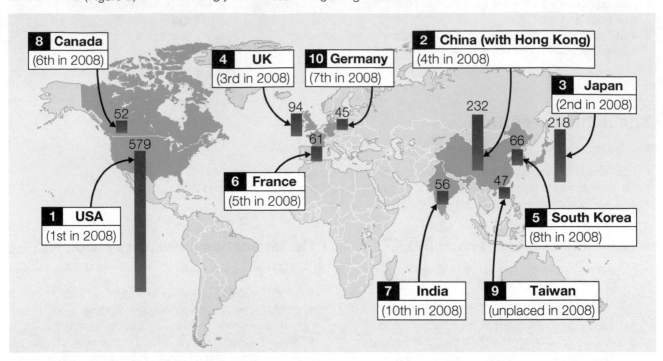

8 Canada (6th in 2008)

4 UK (3rd in 2008)

10 Germany (7th in 2008)

2 China (with Hong Kong) (4th in 2008)

3 Japan (2nd in 2008)

1 USA (1st in 2008)

6 France (5th in 2008)

5 South Korea (8th in 2008)

7 India (10th in 2008)

9 Taiwan (unplaced in 2008)

52   579   94   45   61   232   218   66   56   47

Figure 2 *The ten countries where 75% of the world's 2000 largest companies were based in 2015. Each bar is in proportion to the number of large companies in that country*

TNCs as players in global trade

TNCs have had huge impacts on global trade as the global shift in manufacturing has led to rapid increases in exports from developing countries.

- This has shifted economic power towards emerging countries and made TNCs extremely powerful.
- Much of the increase in global trade is *intra-company* (i.e. trade between different branches or activities of the same company).

TNCS as players in technology

The introduction of the Trade-Related Aspects of Intellectual Property Rights (TRIPS) in 1995 (see Section 3.8) requires all WTO members to protect and enforce their **intellectual property rights**.

- Any new technology or process must be registered under **patent** law (laws to grant exclusive right of ownership of intellectual property). Anyone who wishes to use such property must pay a royalty.
- TRIPS favours TNCs as it protects investment into research and innovation. For example, patent developments by pharmaceutical companies make many medicines (e.g. new treatments for HIV) unaffordable to developing countries.

Global cultural influences

The dominance of Western TNCs ensures that Western culture is global.

- *The Simpsons* is an example; it is available almost everywhere, including airlines and streaming channels.
- The same applies to food, e.g. the vast range of international foods available in UK supermarkets.

For more information and examples of cultural diffusion, see page 177 of the Year 1 student book.

 Ten-second summary

- Superpowers influence the global economy through a variety of IGOs, including the WEF, IMF, WTO and the World Bank.
- TNCs are major players in the global economy.
- TNCs' influence extends to global trade, technology, patents and global culture.

 Over to you

Referring to pages 118–120 of the student book, list the similarities and differences between the WTO, WEF, IMF and World Bank.

You need to know:
- that superpowers and emerging nations play a key role in global action
- how military, economic and environmental alliances increase interdependence and have global influence
- that the UN is important to global geopolitical stability.

Big idea
Superpowers and emerging nations play a key role in international decision-making.

Responses to crisis

Superpowers and emerging nations respond to crises, e.g. when hurricanes and earthquakes hit. Developing nations look for help from other countries, ranging from financial aid to materials for rebuilding. When Hurricane Matthew hit Haiti in 2016:

- Over 900 Haitians died, several thousand homes were destroyed and 350 000 Haitians needed aid.
- The USA and France deployed 550 personnel and provided humanitarian aid.
- NGOs (e.g. the Red Cross) and IGOs (e.g. UNICEF) launched appeals for funds.

Responses to conflict

Superpowers often act as 'global police' in military conflicts, e.g. in Afghanistan (Figure 1).

- **Geopolitics** and religious tensions have led to conflict in Afghanistan for 40 years.
- It is strategically important for trade routes into central Asia and Russia.
- Its mountainous terrain makes central government control difficult.

- Following 9/11, the USA led an international military coalition against the Afghan Taliban, seeking to destroy training camps and kill militant leaders.
- The USA's role and presence antagonised supporters of reform in Afghanistan.
- Despite the killing of Taliban leader Osama bin Laden in 2011, militants continue to attack forces and government targets.

Russia no longer has a direct border with Afghanistan. **Uzbekistan, Turkmenistan** and **Tajikistan** (part of the USSR until 1991) do and they are concerned about the rise of Islamist groups in neighbouring countries.

China is not too concerned about Afghanistan. But, with a large Muslim population in western China, it fears any destabilising Islamist threats within Asia.

Since an Islamist attack on a Mumbai hotel in 2008, **India** fears instability from fundamental Islamists based in Afghanistan and Pakistan. India wants a stable, Western-supported government in Afghanistan.

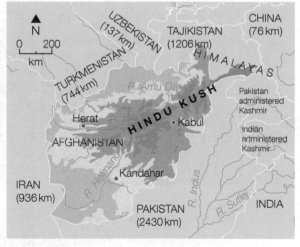

 Figure 1 *The geopolitical location of Afghanistan, showing the complexity of its relationships, and length of the borders, with its neighbours*

For more detail about the conflict in Afghanistan, see page 125 of the student book.

Responses to climate change

Superpowers and emerging nations also play a key role in global action against climate change.

- As impacts increase (e.g. the disappearance of low-lying Pacific islands like Kiribati), the world's first environmental refugees are being forced to leave their homes.

- People who have not yet left Kiribati are taking action to minimise the risk of flooding to their homes. Many have moved away from the coast, rebuilding their wooden houses further inland.
- Those who remain are being squeezed into an ever thinner strip of higher ground.
- As major emitters and contributors to climate change, superpowers are under pressure to act (see Section 3.7).

Attitudes and actions of IGOs towards geopolitical stability

The UN was formed to maintain peace after the Second World War and its sub-organisations are still important to geopolitical stability. These organisations now have 193 members, which contribute to:

- maintaining peace
- promoting human rights
- social and economic development
- providing humanitarian relief and aid work.

International Court of Justice

- It settles disputes between UN member countries and advises on international law.
- It has 15 judges, who represent different global regions.
- It deals with cases brought by individual countries, not by individual people.

UN Security Council and Peacekeeping

- It is ultimately responsible for preventing conflict.
- It has five permanent members, which can veto any resolution.
- It has authorised military and peacekeeping missions to conflicts, e.g. in the Democratic Republic of Congo (2004–2009).

UN and climate change

- The annual UN Climate Conferences aim to make progress in managing climate change.
- The **2015 Paris Agreement** engages all countries in significantly reducing emissions.
- The agreement has been undermined by climate change deniers, such as US President Trump.

International players and global policing

Increasingly, formalised alliances between countries are important in increasing interdependence, geo-strategy and global influence.

Economic alliances

The **European Union (EU)** forms a free-trade area, with 28 member states in 2016.

- Its remit is increasing political union (e.g. free movement of people, a common currency).
- Its guiding principle is that economic strength insures against poverty and policies should work towards reducing inequality.
- Its influence extends into environmental issues and human rights.
- Its future is less stable following the UK vote in 2016 to leave the EU.

For information about NAFTA and ASEAN, see page 129 of the student book.

Military alliances

The **North Atlantic Treaty Organisation (NATO)** was formed in 1949 at the start of the Cold War.

- It remains one of the dominant international military alliances.
- Its guiding principle is that an attack on one member is an attack on all.
- NATO's influence diminished at the end of the Cold War but Russian military activity has brought it to the fore again.

For information about ANZUS, see page 128 of the student book.

Environmental alliances

The **Intergovernmental Panel on Climate Change (IPCC)** was established in 1988 by the UN.

- Its members represent over 120 countries.
- It produces reports on climate change, aiming to assemble evidence from **peer-reviewed** publications to ensure the stabilisation of greenhouse gas concentrations.

Ten-second summary

- Superpowers and emerging nations have a key role in responding to crises, military conflict and climate change.
- The UN and its sub-organisations are important to geopolitical stability.
- Military, economic and environmental alliances increase interdependence and are important in geo-strategy and global influence.

Over to you

Construct a table of strengths and weaknesses for each of the major alliances mentioned in this section. Refer, where relevant, to changing status in the 21st century.

You need to know:

- how superpowers' demand for resources is rising and impacting on the environment
- how growth of middle-class consumption has implications for resource availability and cost
- how superpower growth impacts on carbon emissions and how superpowers differ in their willingness to reduce emissions.

Big idea

Levels of CO_2 emissions and threats to global resources are disproportionately influenced by the actions of superpowers.

Global demand for commodities

As superpowers' demand for resources increases, so their environmental footprint also grows.

- China and the USA lead the way in their demand for most key commodities.
- Each of the global superpowers consumed vast amounts of resources in 2012 (Figure 1).

Impact of China's growth

- China's construction boom makes it the biggest global producer and consumer of steel.
- The impact of China's growth on commodity prices is huge. At the peak of China's growth (2008–2010), copper prices doubled.

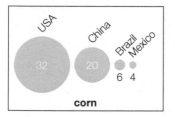

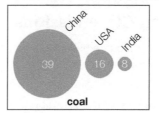

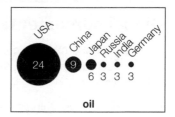

Figure 1 *The main consumers of key commodities in 2012 (as percentages of total global consumption)*

The emerging middle class

Emerging economies have led to increased resource consumption, which has exceeded global population growth since 2000.

- The increased wealth of superpowers has led to more disposable income and resource consumption (Figure 2).
- The 'global middle class' grew by 500 million between 2000 and 2014. Key growth economies are Latin America, South and South-East Asia.
- It is estimated that every 10% increase in a nation's middle class produces an extra 0.5% annual increase in economic growth, driven by demand for resources.

Implications for resources in short supply

The increased demand for resources creates huge environmental impacts and implications. Mobile phones, for example, require:

- crude oil for plastics
- metals, including copper, gold, nickel and zinc
- several toxic and rare compounds for the batteries
- plastic, glass and mercury for the displays.

Figure 2 *Since 2000, China has increased its demand for many commodities*

China's demand for water

One consequence of China's economic growth and demand for resources is water scarcity.

- Its water resources are unevenly distributed and heavily consumed by farming and the coal industry.

- These industries are located in northern China where water is scarce; average water availability per capita is just 200 tonnes.
- In Beijing, total consumption exceeded supply by 70% in 2012, as more residents installed showers and flush toilets.

Monitoring the Earth's atmosphere

The greatest concentrations of CO_2 in the atmosphere (as measured by NASA's monitoring satellite) are found in North America, Europe, India and China (Figure 3). These coincide with the areas of high population density and with developed and emerging economies.

Globally, CO_2 emissions rose by 53% between 1990 and 2013. Within that period, China's emissions increased by 286%, five times the global rate, making it the world's largest CO_2 emitter.

Rank	Country	2013 CO_2 emissions from energy consumption	
		Total CO_2 emissions (Gt – rounded)	CO_2 per capita (tonnes)
1	China	9000	6.5
2	USA	5100	17.6
3	Russia	1700	12.6
4	India	1900	1.5
5	Japan	1200	9.3
6	Germany	750	9.2
7	South Korea	610	12.5
8	Canada	560	16.2
9	Iran	550	8.0
10	Saudi Arabia	510	19.7

 Figure 3 *The ten countries with the highest CO_2 emissions in 2013, by country and per capita. Note the dominance of the global superpowers*

Global agreements on CO_2 emissions

Despite their enormous impact on resource use and greenhouse gas emissions, the superpowers differ in their willingness to act.

- As a key player, China must be included in any global reduction agreement if it is to be successful (Figure 4). In 2016, China agreed to some emissions targets and its progress towards achieving these will determine the success or failure of the 2015 Paris Agreement.
- Russia also supported the 2015 Paris Agreement by agreeing to lower its CO_2 emissions relative to 1990 levels. However, this could actually allow Russia to increase its emissions because its levels in 1990 were much higher than levels after its economic collapse in 1991.
- Since 2005, the USA has led the way, reducing its total emissions by using renewable energy and energy efficiency measures. Whether this will continue under the leadership of President Trump is unknown.
- The EU is another leader of climate initiatives such as carbon trading, emissions reductions and renewable energy. It aims to cut 20% of its energy consumption by 2020.

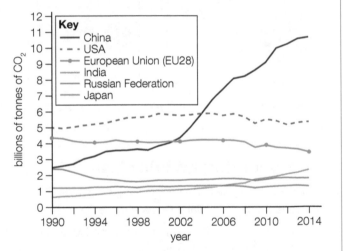

Key
— China
- - - USA
—•— European Union (EU28)
— India
— Russian Federation
— Japan

 Figure 4 *CO_2 emissions from fossil fuel use and cement production in the top five emitting countries and the EU, 1990–2014*

⏱ **Ten-second summary**

- Emerging economies have led to increased resource consumption.
- The growth of middle-class consumption in emerging superpowers has implications for resources.
- The greatest concentrations of atmospheric CO_2 coincide with areas of high population density, and with developed and emerging economies.
- Superpowers differ in their willingness to act to reduce CO_2 emissions.

✏ **Over to you**

Create a mind map to show how rising demand for food, fossil fuels and minerals in China has environmental impacts.

You need to know:

- that tensions can arise over physical resources and territory
- economic tensions can arise over intellectual property rights.

Big idea

The global influence of superpowers is contested economically, environmentally and politically.

Tension over resources

The Arctic (Figure 1) is a good example of where claims have been made to territory and resources there, such as oil and gas there. This happens where:

- ownership of physical resources is disputed
- disagreement exists over their exploitation.

The mineral wealth of oil and gas will become increasingly accessible as, due to climate change, the Arctic Ocean ice thaws more each summer and important new transport routes open up.

This causes political and military tension in the region.

- Since 2002, Canada has carried out military exercises in the Arctic, Norway has expanded its navy, and Denmark has created a military command and response force for the region.
- In 2007, Russia staked a claim to the ownership of resources at the North Pole and carries out Arctic bomber patrols.
- In 2014, Denmark used UNCLOS (the UN Convection on the Law of the Sea) to claim a section of the Arctic.

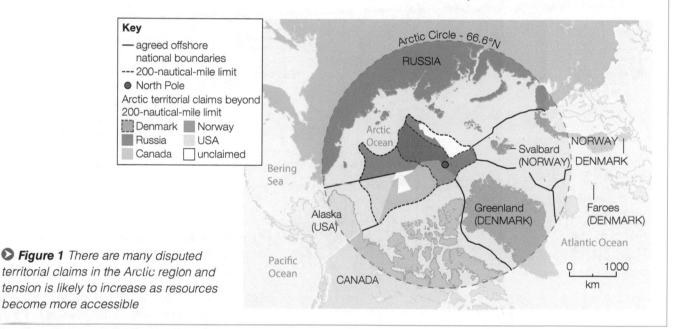

Key
— agreed offshore national boundaries
--- 200-nautical-mile limit
● North Pole
Arctic territorial claims beyond 200-nautical-mile limit
- Denmark
- Russia
- Canada
- Norway
- USA
- unclaimed

▶ **Figure 1** *There are many disputed territorial claims in the Arctic region and tension is likely to increase as resources become more accessible*

Attitudes to resources

Beyond the Arctic region, disagreement over the exploitation of physical resources (e.g. water, land, minerals) can cause tension.

- In the UK, owning land does not necessarily mean owning the mineral wealth that lies beneath it, which results in disputes.
- Iraq's oil wealth is disputed between Sunni, Shia and Kurdish regions.

Economic tensions over intellectual property

Counterfeit products are sold on market stalls around the world, straining global trade relations and threatening TNC investment.

- Faking brands is illegal and an international crime against **intellectual property rights** (IPR) under WTO rules.
- International trade agreements aim to protect brand names. Agreement is known as **Trade-Related Aspects of Intellectual Property Rights** (TRIPS).

The TRIPS agreement incorporates IPR law into international trade and means that WTO members guarantee copyright protection for everything from performance rights to designs.

- It protects **patents** so that new innovations cannot be copied or pirated.
- This global system is undermined by counterfeiting.

Tensions over territory

Political influence often leads to tensions over territory and resources, for example in Eastern Europe (Figure 2).

- The extent of Russia's influence in many Eastern European states has changed over time (Figure 3).
- Tension is growing between Russia and many Eastern European countries, e.g. Poland and some former states of the USSR.
- Meanwhile, the influence of the EU may expand in the region in the future.

Russia has felt its potential loss of influence greatly and, since 2014, has used both hard and soft power to re-establish itself.

- In 2014, it annexed (took control of) Crimea, and has supported separatists fighting in eastern Ukraine.
- It has undertaken combat exercises in the Arctic for the first time since 1991 and is expanding naval and weapons bases there.

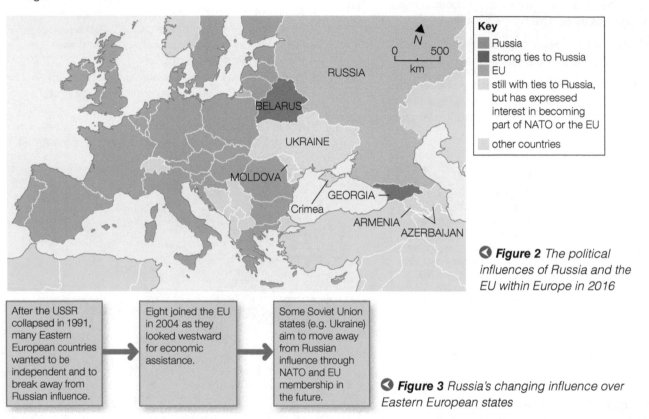

Key
- Russia
- strong ties to Russia
- EU
- still with ties to Russia, but has expressed interest in becoming part of NATO or the EU
- other countries

◄ **Figure 2** *The political influences of Russia and the EU within Europe in 2016*

| After the USSR collapsed in 1991, many Eastern European countries wanted to be independent and to break away from Russian influence. | → | Eight joined the EU in 2004 as they looked westward for economic assistance. | → | Some Soviet Union states (e.g. Ukraine) aim to move away from Russian influence through NATO and EU membership in the future. |

◄ **Figure 3** *Russia's changing influence over Eastern European states*

And as for China?

Until 2010, China's influence over global affairs tended to be economic. That has begun to change.

- It seeks wider influence within Asia and will not abandon long-held territorial claims over Tibet and Taiwan.
- It has used the argument of the extent of its Exclusive Economic Zone to maintain contentious territorial claims in the South China Sea, which conflict with claims of neighbouring countries.

- It has responded to these disputes by creating military bases in the Spratly Islands. This expansion of hard power worries countries such as India and US allies.
- These ambitions are relative, however. China is not involved in conflicts in Syria or elsewhere, and it is an important member of UN peacekeeping missions. The influence it seeks is, for now, confined to Asia.

For a map of disputed territories in the South China Sea, see Figure 7 on page 137 of the student book.

Ten-second summary

Tensions can arise:

- over physical resources, where ownership is disputed and/or disagreement exists over their exploitation.
- over intellectual property rights when undermined by counterfeiting
- over territory.

Over to you

Annotate a map to show how influence is contested in each of the following:

a the Arctic region
b the South China Sea.

You need to know:

- developing economic ties between emerging powers and the developing world bring benefits and problems
- the rise of China and India is creating tensions in Asia
- tensions in the Middle East represent an on-going challenge.

Big idea

Developing nations have changing relationships with superpowers, with consequences for people and the environment.

China in Africa

China's economic growth has increased its demand for resources, so it has increased its trade relations with the developing world, particularly Africa (Figure 1).

- Its involvement with Africa has focused on trade and investment in infrastructure, e.g. transport links such as railways for exporting raw materials.
- By 2015, China had become Africa's largest trading partner, leading to US$60 billion in Chinese Foreign Direct Investment (FDI), boosting mining, agriculture, banking and IT.
- Over 1 million Chinese (mostly labourers and traders) have moved to Africa since 2005.

Environmental impacts in Africa

Despite providing employment, China's decision to move some industries to Africa (e.g. steel and cement) has caused concern owing to air and water pollution.

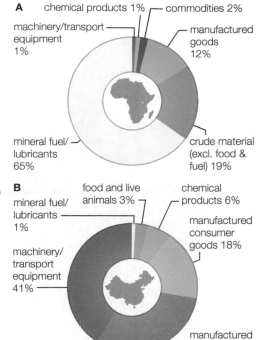

> **Figure 1** *Trade flows between Africa and China in 2015:* **A** *shows African exports to China and* **B** *shows Chinese exports to Africa*

Growing tensions within Asia

The rise of China and India brings political tensions and could affect geopolitics in Asia.

Taiwan

The Chinese civil war established the communist People's Republic of China, with the opposition government retreating to Taiwan.

- Since 1949, there have been tensions between the two, each claiming to be China's legitimate government.
- Communist China maintains its territorial claim to Taiwan as a Chinese province.

Tibet

The diaspora (people living without a home territory) in Tibet seek political separation from China.

- One reason is religious differences with Beijing's atheist government refusing to acknowledge the Dalai Lama as the traditional spiritual and political leader of Tibet.
- By encouraging ethnic Chinese migrants to move to Tibet, China has raised tensions.

Japan

- Despite long-standing tensions between China and Japan, they are now major trading partners.
- Historic tensions stem from the stationing of US troops in Japan during the Chinese civil war and the adoption of a capitalist, westernised economy by Japan following the Second World War.

India

Tense relations between India and China have been based on border disputes.

- Both maintain a military presence along their Himalayan borders and India is suspicious of China's good relationship with Pakistan.
- Equally, China is concerned about India's military interest in the South China Sea.

Key players: the role of emerging powers

Emerging economies are playing an expanding role on the world stage. Globalisation has altered the geopolitical power and influence of the Asia region.

- By 2050, the economic centre of gravity is likely to be between India and China, having shifted from the mid-Atlantic in 1980 (see Section 3.2).
- There is an increasing reversal of FDI flows from emerging to developed countries.

- In 2013, China proposed its 'One Belt, One Road' strategy to develop connectivity between China, Eurasia and Africa (Figure 2). The plan to expand Chinese influence westwards involves infrastructure to increase trade and cultural exchanges. China imagines this as forming a single cohesive economic area, putting it firmly on the world stage.

◀ *Figure 2* China's 'One Belt, One Road' strategy

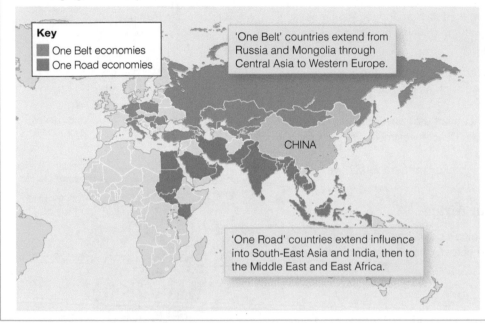

Key
- One Belt economies
- One Road economies

'One Belt' countries extend from Russia and Mongolia through Central Asia to Western Europe.

CHINA

'One Road' countries extend influence into South-East Asia and India, then to the Middle East and East Africa.

Attitudes and actions: contrasting cultural ideologies

Tensions and trouble spots in the Middle East present a challenge to superpowers and emerging powers because of complex geopolitical relations and the need to maintain oil supplies. Volatile swings in global oil prices occur following threats of conflicts.

Political tensions

Israel is a contentious state with the Muslim Middle East, which opposes its creation. Therefore its major supporter (the USA) struggles to maintain a positive influence in the Arab world.

Cultural tensions

This includes religious and ethnic tensions stemming from historic tribal or religious divisions of the region, e.g. between Shia Iran and Sunni Iraq.

Economic tensions

The region is an essential supplier of oil. The rise of ISIS in Iraq has focused Western interests on defending the country's oil reserves.

Environmental tensions

Past conflicts have often resulted in damage to oil installations with environmental consequences, e.g. following the invasion of Allied forces in Iraq in 2003.

 Ten-second summary

- China's economic growth has led to greater ties with Africa.
- The rise of China and India brings political tensions.
- Trouble spots in the Middle East present a challenge to superpowers and emerging powers.

 Over to you

1 Construct a table that shows the positives and negatives of China's investment in Africa.
2 Draw a spider diagram to outline tensions facing the Middle East. Then identify which of these you regard as most serious.

You need to know:

- that economic problems represent an on-going challenge to the USA and EU
- that the costs of maintaining global military power and space exploration are being questioned
- that the future balance of global power is uncertain.

Big idea

Existing superpowers face on-going economic problems that challenge their power.

Economic restructuring

Economic restructuring in Western countries shifted employment from manufacturing into tertiary and quaternary sectors. This created long-term challenges.

- **Economic costs** – widespread **unemployment**, with the loss of traditional mining and manufacturing industries (Figure 1).
- **Social costs** – social cohesion was lost because of a spiral of decline (e.g. where people lost full-time work and faced mental health problems) and because people often had to migrate to find work.

However, the growth in tertiary and quaternary sectors is not without problems, as the financial crisis of 2008 showed.

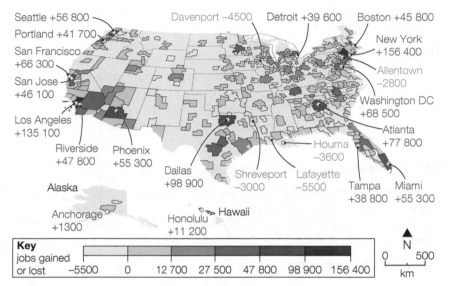

Seattle +56 800
Portland +41 700
San Francisco +66 300
San Jose +46 100
Los Angeles +135 100
Riverside +47 800
Phoenix +55 300
Alaska
Anchorage +1300
Davenport −4500
Detroit +39 600
Boston +45 800
New York +156 400
Allentown −2800
Washington DC +68 500
Atlanta +77 800
Houma −3600
Dallas +98 900
Shreveport −3000
Lafayette −5500
Tampa +38 800
Miami +55 300
Honolulu +11 200
Hawaii

Key jobs gained or lost: −5500   0   12 700   27 500   47 800   98 900   156 400

N   0   500 km

▲ **Figure 1** *Changing employment in the US, 2015. Tertiary and quaternary employment have grown in New York, Los Angeles and 'sun-belt' regions, while traditional manufacturing cities have declined*

The debt problem and financial crisis

The global debt crisis of 2007–2008 arose largely from US and European mortgage-lending markets.

- **Sub-prime lending** by US banks allowed low-income earners with insecure jobs to be given mortgages that they would struggle to repay. These risky mortgages were packaged with more secure investments and sold on to global banks.
- This led to banks having debt that became worthless when the property bubble burst.
- Owing to the globalisation of the banking system, the collapse affected every global bank.
- Confidence in the global banking system was shattered as some of the world's largest banks collapsed (e.g. Lehmann Brothers, a US investment bank).
- In the UK, some banks (e.g. Lloyds) were bailed out with government money, while others (e.g. Royal Bank of Scotland, Northern Rock) were **nationalised**.
- Bank bailouts increased the debt levels of many governments (e.g. the UK and USA).

Impacts of the financial crisis

Despite bank bailouts, the repercussions of the financial crisis were huge and countries reacted differently.

- The US government increased national debt to maintain consumer spending and an economic multiplier effect.
- The UK government adopted **austerity** and reduced government spending, creating flat or little growth.
- Greece was one of the worst hit countries where high unemployment led to hate crimes against immigrants.

Maintaining global military power

As a uni-polar superpower with vast military spending, countries look to the USA (and its allies) for help and military assistance when tensions rise. Whether the USA and its allies should, or can afford to, keep bearing these costs in the future is now being debated.

National defence budgets are usually large. In 2016/17, the UK had the world's fifth largest defence budget (US$56 billion) – 10% of the US defence budget.

Naval power

- Austerity led to a decline in UK government defence spending.
- Current debates centre around whether it is better to have a larger number of low-cost ships (strength in numbers) or a smaller number of high-tech ships.

Nuclear weapons

- The on-going debate is whether to invest in weapons that may never be used.
- Weaponry can deter the escalation of conflict, yet some argue that the main beneficiaries are US defence industries.
- Despite considerable cost, in 2016 the UK Parliament voted to replace the UK's Trident nuclear deterrent.

Air power

- The research and development needed to improve and maintain combat aircraft can increase their cost sharply.
- Current warfare is focused on rapid response air power (Figure 2), as opposed to naval fleets.

Intelligence services

- Government budgets are increasingly directed towards anti-terrorism work.
- This is labour-intensive (involving intelligence) and expensive.

Space exploration

- Space budgets promote exploration (e.g. joint funding of the International Space Station) but are under threat in many Western countries.
- India and China are adopting major space programmes and launching space flights more cheaply.

Country	Number of combat aircraft
USA	3318
Russia	1900
China	1500
India	1080
Egypt	900
North Korea	661
Pakistan	502
Turkey	465
South Korea	458
Germany	423

 Figure 2 *The world's largest air forces in 2016 by the number of fixed-wing combat aircraft (i.e. helicopters are not included). The UK was eleventh*

For information about how NATO helps to maintain global military power, see page 145 of the student book.

Futures and uncertainties: future power structures

The balance of global power in 2030 and 2050 is uncertain, with several possible outcomes:

- continued USA dominance
- bi-polar structures
- multi-polar structures.

Based on current data, China could overtake the USA as the world's largest economy between 2025 and 2030. However, in the mid-1980s, Japan was predicted to challenge the USA's dominance and the Asian financial crisis of 1997 prevented that. So predictions about China might not materialise.

 Ten-second summary

- Economic restructuring creates economic challenges.
- Debt and financial crisis resulted in austerity and high unemployment in parts of Europe.
- The economic costs of maintaining global military power and space exploration are being questioned.
- The future balance of global power is uncertain with several possible outcomes.

 Over to you

Review all your work in 'Over to you' activities in this chapter of the Revision Guide. Weigh up the evidence that by 2050 we will have the following global superpower structures:

- continued USA dominance
- a bi-polar structure
- a multi-polar structure.

Which outcome do you think is most likely and why?

Chapter 4
Health, human rights and intervention

What do you have to know?

This chapter studies how traditional economic definitions of development are challenged by definitions based on quality of life and measures used to record progress in human rights and human welfare. National and global institutions impact on decisions leading to geopolitical intervention, from development aid to military campaigns.

The specification is framed around four enquiry questions:

1 What is human development and why do levels vary from place to place?
2 Why do human rights vary from place to place?
3 How are human rights used as arguments for political and military intervention?
4 What are the outcomes of geopolitical interventions in terms of human development and human rights?

The table below should help you.

- Get to know the key ideas. They are important because 20-mark questions will be based on these.
- Copy the table and complete the key words and phrases by looking at Topic 8A in the specification. Section 8A.1 has been done for you.

Key idea	Key words and phrases you need to know
8A.1 Concepts of human development are complex and contested.	human development, human contentment (Happy Planet Index), Sharia law, quality of environment and health as goals for development, human capital, human rights, gender equality, UNESCO
8A.2 There are notable variations in human health and life expectancy.	
8A.3 Governments and International Government Organisations play a significant role in defining development targets and policies.	
8A.4 Human rights have become important aspects of both international law and international agreements.	
8A.5 There are significant differences between countries in both their definitions and protection of human rights.	
8A.6 There are significant variations in human rights within countries, which are reflected in different levels of social development.	
8A.7 There are different forms of geopolitical intervention in defence of human rights.	
8A.8 Some development is focused on improving both human rights and human welfare but other development has very negative environmental and cultural impacts.	
8A.9 Military aid and both direct and indirect military intervention are frequently justified in terms of human rights.	
8A.10 There are several ways of measuring the success of geopolitical interventions.	
8A.11 Development aid has a mixed record of success.	
8A.12 Military interventions, both direct and indirect, have a mixed record of success.	

You need to know:

- that human development can be measured in different ways
- about the relationship between development, economic growth and human well-being
- an example of an alternative approach to human development
- that access to education is important and varies around the world.

Big idea

Human development is complex and contested.

Geography and development

Development means a change for the better. Over the past 50 years, the global economy has grown. But some countries have improved rapidly and some have improved more slowly.

Professor Hans Rosling was an expert in development data and presented health, wealth and population statistics in a visually engaging way. He thought that:

- Countries should aim to improve environmental quality, health, life expectancy and human rights.
- Economic growth is the most important way of improving these.
- A stable government is necessary to improve human rights.

Measuring human development

There are several ways to measure human development, and its measurement is both complex and contested.

GDP per capita

- This most commonly used measure of wealth is calculated by dividing the value of all goods and services produced within a country by its population.
- It is an average and hides wide differences in income within a country.
- Countries with similar GDP vary in levels of human well-being. For example, Italy and India have similar total GDP (US$ 2.1 trillion) but quite different HDI (Italy = 0.88 and India = 0.62).

Gini coefficient

- The **Gini coefficient** shows how unequal a country is (Figure 1).
- It is scored between 0 and 1, where 0 = everyone has the same income (perfect equality) and 1 = one person has all the income (greatest inequality).
- Generally, poorer countries are most unequal.

Human Development Index (HDI)

- The **HDI** uses four different measurements to create a single index figure.
- It is based on measurements of life expectancy, education and GDP.
- It scores human development between 0 and 1 (0 = low, 1 = high).

Happy Planet Index (HPI)

- The **HPI** measures happiness and sustainable human well-being.
- It uses measurements of **life expectancy**, **experienced well-being** and **ecological footprint**.
- It scores well-being between 0 and 100 (0 = low, 100 = high).

Figure 1 ▸

The Gini coefficient for a selection of HICs and MICs (2011). HICs can also have varying levels of inequality

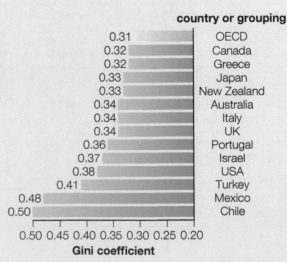

country or grouping

Gini coefficient	Country or grouping
0.31	OECD
0.32	Canada
0.32	Greece
0.33	Japan
0.33	New Zealand
0.34	Australia
0.34	Italy
0.34	UK
0.36	Portugal
0.37	Israel
0.38	USA
0.41	Turkey
0.48	Mexico
0.50	Chile

0.50 0.45 0.40 0.35 0.30 0.25 0.20
Gini coefficient

Alternative ways of assessing human development

There are alternative approaches to assessing improvement in human development. For example:

- **Human welfare** – Sharia law is the law of Islam. Muslims believe that the welfare of humans is based on the fulfilment of necessities, needs and comforts.
- **Intervention by national government** – Political changes, such as those in Bolivia under President Evo Morales, can alleviate poverty and lead to economic growth, in this case, through taking control of natural resources.

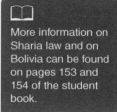

More information on Sharia law and on Bolivia can be found on pages 153 and 154 of the student book.

Education matters

Investment in education and health is seen as an investment in **human capital** (economic, political, cultural and social skills).

- Education is also key to improving human rights and democratic participation.
- The UN recognises the right for free primary education for everyone because it allows '*full development of the human personality*'. However, 60 million children still do not attend primary school (Figure 2), 32 million in sub-Saharan Africa.

- There is a **gender imbalance** as girls make up 54% of the world's non-schooled population, particularly in Asia and the Arab states.
- The educational level children have reached when they leave school varies (Figure 3).
- A lack of education can slow down the social and economic development of a country.

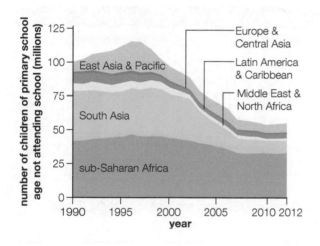

Figure 2 *The number of children not attending primary school has fallen, but varies between different world regions*

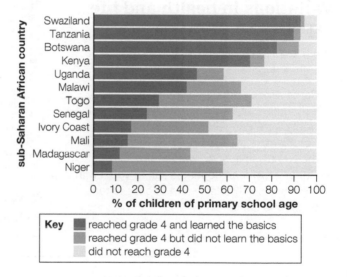

Figure 3 *The percentage of children reaching grade 4 (basic literacy and numeracy) in some sub-Saharan African countries, 2013/14*

 Ten-second summary

- Economic growth can help to improve environmental quality, health, life expectancy and human rights.
- Development can be measured using different indicators, e.g. GDP, Gini coefficient, HDI and HPI.
- There are alternatives to accepted models of human development.
- Education is key to economic development and improving human rights, but access to it varies globally, particularly for girls.

Over to you

Draw a mind map with 'Development' as the central word. Add all of the key ideas in this section and explain how each idea is connected to development.

You need to know:

- that health and life expectancy can vary between and within countries
- that variations in the developing world are due to access to food, water and sanitation
- that variations in the developed world are due to lifestyle choice, deprivation and healthcare
- that variations can be due to income levels and ethnicity.

Big idea

There are variations in health and life expectancy.

Factors affecting health and life expectancy

Several factors strongly influence health and life expectancy, including differences in lifestyles, levels of deprivation and access to healthcare (Figure 1). These variations occur:

- across the developing world
- across the developed world
- within countries.

Wider determinants
- Occupation
- Education
- Income

Preventative healthcare
- Immunisation

Lifestyle factors
- Smoking
- Diet
- Alcohol

▶ **Figure 1** *Several factors (or determinants) influence health and life expectancy*

Variations in health and life expectancy in Africa

There are wide variations between African countries (Figure 2).

- Some African countries improved their economies and the lives of people when their debts were cancelled in 2005.
- For example, Kenya and Tanzania had two of the world's fastest growing economies in 2016.
- But other African countries still struggle with high levels of mortality and access to food, water and sanitation.

Democratic Republic of the Congo	Algeria
• US$800 GDP per capita (PPP) • Life expectancy – 56 years • One of the world's lowest HDI • Rich in natural resources, e.g gold • However, resources have caused conflict and up to 6 million deaths • Has met few of the **Millenium Development Goals (MDGs)** • 40% of children are forced to work • High health expenditure and funding from NGOs	• Rapidly increasing GDP, mainly from oil • Life expectancy – 76 years • HDI of 0.736, one of Africa's highest • Strong government • Primary school enrolment of 98.16% • Has met most of the Millenium Development Goals (MDGs) • Still has 23% of the population living below the poverty line

▲ **Figure 2** *Comparing Democratic Republic of the Congo and Algeria*

Variations in health and life expectancy in the developed world

Between 1985 and 2010, the Organisation for Economic Co-operation and Development (OECD) saw health and life expectancy continually improving in its 35 member nations from the developed world. Their research found that:

- In developed countries, life expectancy rose.
- Infant mortality rates and deaths from heart disease fell.
- Rising spending on healthcare, better medicine, reducing smoking and calorie intake has helped to improve health.
- The USA has the highest health spending per capita in the world (Figure 3), but only ranks 30th for infant mortality rate.
- Healthcare provisions vary, e.g. the USA's private sector and the UK's state-funded NHS.

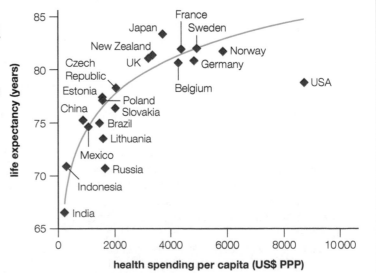

▲ **Figure 3** *Life expectancy at birth and health spending per capita for selected OECD countries in 2015*

Variations within the UK

The UK's average life expectancy is 82.8 for women and 79.1 for men. It is expected to continue increasing. But there are regional variations across the country (Figure 4).

UK average life expectancy is rising overall.

North East England has below average life expectancy.

For men and women at age 65, life expectancy is highest in London.

Spendng on fresh, healthy food is higher in the South East.

English region	Average life expectancy (male)	Average life expectancy (female)
North East	77.8	81.6
North West	77.7	81.7
West Midlands	78.7	82.7
London	79.7	83.8
South East	80.3	83.8
South West	80.0	83.9

Many deaths are linked to alcohol and smoking.

For women at age 65, life expectancy is lower in the North West.

⬤ **Figure 4** *There are regional variations in life expectancy within the UK*

Variations within Australia

Australia has one of the highest life expectancies in the world.

- It is one of only seven countries with average life expectancies of over 80 years for men and women.
- But life expectancy for the indigenous Australians is around 10 years lower than for non-indigenous Australians (Figure 5).
- Some social indicators for indigenous Australians are as low as some sub-Saharan African countries.
- Indigenous Australians have higher levels of deprivation, smoking, disease and injury, as well as lower levels of education and employment.
- This means that regions with a high proportion of indigenous people have lower life expectancies.

In 2009, the Australian government launched 'Closing the Gap' to improve health, education and life expectancy for indigenous people.

📖 More information on the UK and Australia can be found on pages 160 and 161 of the student book.

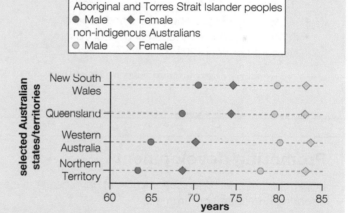

Key
Aboriginal and Torres Strait Islander peoples
- ⬤ Male   ◆ Female
non-indigenous Australians
- ○ Male   ◇ Female

⬤ **Figure 5** *Variations in average life expectancy between Aboriginal and Torres Strait Islander peoples and non-indigenous Australians, by gender and by selected states and territories (2010-12)*

Ten-second summary

- Several factors impact upon health and life expectancy, e.g. occupation, income and lifestyle choices.
- There is variation in health and life expectancy between African countries, with some improving faster than others.
- Health and life expectancy is improving in the developed world due to rising spending on healthcare.
- There are variations in health and life expectancy within countries.

Over to you

In 20 seconds, try to explain to a friend the reasons for variations in health and life expectancy:

a between African countries
b between developed (OECD) countries
c within the UK
d within Australia.

You need to know:

- that the relationship between economic and social development depends on decisions made by governments
- about the aims and recent programmes of IGOs (IMF, World Bank and WTO) in terms of development
- that there has been mixed progress towards the Millennium Development Goals (MDGs)
- that Sustainable Development Goals (SDGs) build on the successes of the MDGs.

Big idea

Governments and IGOs have a very important role in setting development targets and policies.

Economic and social development

Economic development in a country does not always lead to social development. It depends on the decisions made by governments and the areas they choose to spend money on. There can be wide variations in spending on **healthcare**, **welfare**, **pensions** and **education**.

France

- Has one of the highest levels of government spending – 56% of GDP.
- The majority of healthcare is state-funded.
- It has high welfare and pension payments.
- It has high education spending – £8500 per student per year (2015).

Saudi Arabia

- Is ruled by the elite royal family.
- The healthcare system is high quality and 80% state-funded.
- Welfare and pension spending varies.
- Education standards are low, so skilled and managerial jobs are taken by overseas employees.
- There are high levels of youth unemployment and poverty levels.

Promoting development

Inter-governmental organisations (IGOs) have been criticised for previously adopting neo-liberal programmes.

See pages 118–120 of the student book for background information on the IMF, World Bank and WTO and pages 164–165 for further examples of their programmes.

- These intervened in the policies of individual governments, with the aim of improving economic growth so that wealth would trickle down to the poorest people.
- Countries were forced to promote free trade, privatise government services and assets, and remove barriers to private investment.
- This had the effect of cutting education and healthcare programmes.

More recently, IGOs have focused on programmes with a more direct impact on improving the lives of people (Figure 1).

	World Bank	IMF	WTO
Aim	• To finance loans for development	• To strengthen weak currencies and promote economic development	• To encourage trade as a way of promoting economic development
Examples of programmes	• **Global Partnership for Education**, 2002 • Develops early reading and numeracy skills • Focuses on disadvantaged children (girls, ethnic minorities, disabled) • Invested over US$35 billion in educational programmes	• **Poverty Reduction Programme**, 2000 • Helps countries to create their own development plans • Countries receive aid, loans and debt relief in return	• They restrict the trade of endangered products • They challenge trade that may have a negative environmental impact, such as forest clearance
Criticisms	• Previously financed projects with costly repayments and poor environmental consequences	• Previous policies led to reduced spending on healthcare and education	• Previous policies encouraged countries to damage the environment

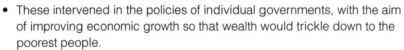

Figure 1 *World Bank, IMF and WTO programmes are significant in promoting development*

Achievements of the Millennium Development Goals

Advances have been made towards the eight Millennium Development Goals (MDGs) introduced by the UN in 2000 (Figure 2). But progress has varied around the world (Figure 3). China's economic progress accounts for over 500 million of the fall in numbers living in extreme poverty. Those who are disadvantaged (due to gender, age, ethnicity, disability) have benefited the least.

Number living in extreme poverty has fallen by 56%, from 1.9 billion in 1990 to 836 million in 2015.

Rate of child mortality has declined by more han 50%, from 90 to 43 deaths per 100 live births since 1990.

Improvements in attendance of girls at primary school.

Over 6.2 million malaria deaths avoided.

Increase in the number of women in parliament in 90% of countries.

2.1 billion people have improved sanitation.

▶ **Figure 2** *Global progress towards the MDGs by 2015*

▶ **Figure 3** *Global progress in achieving MDG Goal 1 (reducing the percentage of people living on less than US$1.25 a day)*

Progress in Bangladesh

Bangladesh had the world's third largest number of people in poverty (33% below the poverty line). But an average 6% GDP growth rate meant that by 2015 Bangladesh had achieved:

- 97.7% of children in primary education, with equal numbers of girls and boys
- reductions in infant and maternal mortality
- reduction in the prevalence of diseases, such as HIV/AIDS and malaria.

The MDGs can be found on page 164 of the student book.

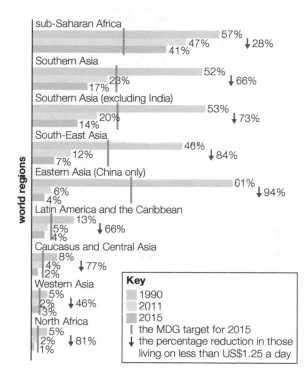

The 2015 Sustainable Development Goals

In 2015, the UN launched a set of 17 **Sustainable Development Goals (SDGs)** to build on the MDGs. These goals:

- aim to end poverty, protect the planet and ensure prosperity for all, by 2030
- focus on strategies to grow economically by addressing health, education and employment needs, whilst tackling climate change.

However, the SDGs are not legally binding.

 Ten-second summary

- The amount of government money spent on health, welfare and education vary between countries.
- Previous IGO policies have been criticised, so more recent policies focus on improving health, education and the environment for people.
- Progress towards the MDGs has been variable, with some countries developing more than others.
- The SDGs build on the MDGs and are part of a development agenda.

Over to you

Which of the MDGs is most important for economic and social development? Rank the eight MDGs on page 164 of the student book in order. Make sure you can explain your rank order.

Student Book
See pages 168–173

You need to know:

- that the Universal Declaration of Human Rights (UDHR) outlines the rights that every human being is entitled to
- that the European Convention on Human Rights (ECHR) protects human rights and is incorporated into British law
- that the Geneva Conventions protect those involved in armed conflict.

Big idea

Human rights are important factors of international law and agreements.

The European Convention on Human Rights (ECHR)

The ECHR is a treaty produced by the Council of Europe in 1950 to protect human rights.

- It consists of 14 articles, which protect rights such as the rights to life, a fair trial and freedom of expression.
- It is now included within the laws of 47 European countries.
- This enables human rights cases to be heard within the home country rather than in the European Court.
- It was integrated into British law as part of the 1998 Human Rights Act.

The Universal Declaration of Human Rights (UDHR)

Produced by the UN after the Second World War, the UDHR outlines the rights that every human being should be entitled to.

- Signed in 1948, it formed a basis for freedom, justice and peace. It lists 30 articles, which define basic human rights.
- It is not legally binding but it was the basis of the International Covenant on Civil and Political Rights and the International Covenant on Economic, Social and Cultural Rights, which provide a legal way to enforce the UDHR.
- 48 countries signed in 1948; those that chose not to include South Africa, whose system of apartheid violated many of the rights outlined in the UDHR.

The 1998 Human Rights Act

This Act incorporates the rights of the ECHR into British law and applies to every UK resident. It has strengths:

✓ British decisions can be contested against a wider frame of rights.
✓ Human rights cases can be heard in British courts, rather than in the European Court of Human Rights.
✓ It requires all public bodies and private organisations to treat everyone equally.
✓ It ensures all new laws passed by Parliament are compatible with the ECHR.

Some people are unhappy that it removes the right of the British Parliament to determine its own laws.

✗ British courts are also bound by decisions made by the European Court.
✗ The European Court ruled that the UK has been in violation of the ECHR in 60% of the cases brought before it (Figure 1).
✗ There are plans to replace the Act with a 'British Bill of Rights and Responsibilities' to return power to British courts.

▶ **Figure 1** *Between 1959 and 2013, in cases brought before it, the ECHR found the UK to be in violation of these nine human rights the most frequently. In 297 of the 499 cases, the judgements found at least one violation*

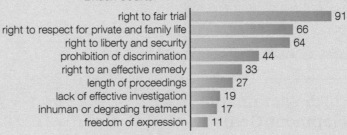

right to fair trial	91
right to respect for private and family life	66
right to liberty and security	64
prohibition of discrimination	44
right to an effective remedy	33
length of proceedings	27
lack of effective investigation	19
inhuman or degrading treatment	17
freedom of expression	11

It's against human rights...

During the 2003–2008 Iraq War, there were allegations that British soldiers subjected Iraqis to harsh interrogation and beatings.

- There has been much debate over whether the UK's 1998 Human Rights Act applies to local people and British armed forces in combat zones.
- However, £20 million was paid in compensation for 326 cases of alleged abuse in Iraq.

The Geneva Conventions

The original Geneva Convention was established in 1864 because of changes in the nature of armed conflict and the need for protection against advanced weaponry.

- After 1945, four Geneva Conventions were created to cover armed forces and civilians involved in conflict.
- They are a body of rules that protect civilians, and those no longer fighting, anywhere in the world.
- Now, almost all countries around the world have ratified the four Conventions. Britain did so in 1957.
- The Conventions are used to determine what counts as a 'war crime'.

Criticisms of the Conventions

- Violations of the Geneva Conventions rarely come to trial.
- 141 countries are reported to use torture still, including the USA.
- Alleged torture of terror suspects occurs at Camp Delta at Guantanamo Bay, Cuba (see Section 4.9).
- Some reporters claim that the Conventions are being abused by all sides in the Syrian War.
- Under the Conventions, emergency service workers in Syria are entitled to protection from attack regardless of their political allegiances. Some news reporters say these workers are not being protected from harm.

Convention rules

1 Injured combatants are entitled to respect for their lives.
2 An enemy who surrenders cannot be harmed.
3 Wounded people shall be cared for.
4 Captured combatants or civilians are entitled to personal rights and dignity.
5 No one shall be subject to torture or degrading treatment.
6 Weapons cannot be used to cause unnecessary loss or suffering.
7 Civilians should not be the object of attack.

Find out more about the war in Syria on page 171 of the student book.

Prosecuting 'war crimes'

The successful prosecution of war crimes is rare.

- There is often little reliable evidence, few witnesses and the circumstances of the crimes can be obscured.
- UN bureaucracy can also hinder the process.
- However, the UN has been successful in prosecuting one of the leading men in the Yugoslav conflict (Figure 2).
- In March 2016, former Bosnia Serb leader, Radovan Karadzic was convicted of genocide, war crimes and crimes against humanity during the 1990s Yugoslav wars. He was found guilty of the Srebrenica massacre, which targeted the entire male Bosnian Muslim population living in the town.

 Figure 2 *Radovan Karadzic was sentenced to 40 years in prison for genocide and war crimes*

Ten-second summary

- The UDHR and the ECHR are important international agreements for the protection of human rights.
- The rights of UK residents are protected by the 1998 Human Rights Act.
- Both the Human Rights Act and the ECHR are criticised by some.
- The rules of the Geneva Conventions protect the human rights of those involved in armed conflict, but many violations are not brought to trial.

Over to you

Make sure that you understand the importance and the role of the four agreements outlined in this section.

You need to know:

- that some countries actively promote and protect human rights, whilst others prioritise economic development
- that even though some countries have grown economically, their levels of human rights differ because of different political systems
- that all countries have some level of corruption, which threatens human rights.

Big idea

There are differences between countries in the way they define and protect human rights.

Economic development versus human rights

Some countries are advocates for human rights and promote their protection in international forums (e.g. USA). However, others prioritise economic development over human rights, e.g. Singapore.

The USA

- It promotes human rights in the UN Human Rights Council, but it only joined the Council in 2009.
- In 2015–2016, it supported resolutions to focus on human rights abuses in Syria, Burundi and Yemen.
- It drew attention to human rights abuses in Cambodia.
- It supported resolutions to ensure the safety of journalists and the rights of indigenous peoples.

Singapore

- It has one of the world's highest GDPs per capita, resulting from its heavy engagement with overseas trade.
- The government argues that preserving order and its relative disregard for human rights has allowed its economic growth.
- It limits rights such as freedom of expression, peaceful assembly and of the press.
- It still uses capital punishment and the death penalty, and has issues around human trafficking.

Democratic freedom for all?

The economies of both China and India have grown rapidly (Figure 1), but by adopting different political systems. Both are now considered to be economic and political superpowers.

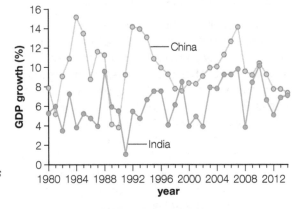

> *Figure 1* The annual economic growth rates for India and China, 1980–2014

Authoritarian government: China

- China is a a single-party authoritarian state, governed by the Chinese Communist Party since 1949.
- There are no general elections.
- The President holds all political power.
- The government limits freedom of expression, association, assembly and religion.
- It argues that human rights are Western ideas and threaten the government's power.
- More recently, greater wealth, better schooling and access to social media have increased calls for more democratic freedom.

Democratic government: India

- Individual Indian states and the Lower House of Parliament hold the most power.
- India's Constitution consists of 444 articles, including freedom of speech and religion.
- The media and independent judiciary ensure the freedom of society.
- Despite significant progress in protecting human rights, concerns over gender inequality, discrimination and disability rights still exist.
- The government is committed to improving human rights as the economy grows.

Measuring political corruption

Transparency International (an international NGO) aims to combat corruption and prevent criminal activities arising from corruption.

- It defines corruption as '*the abuse of entrusted power for private gain*'.
- Corruption often involves decision-makers who abuse their positions to sustain power, status and wealth.

Transparency International publishes a Corruptions Perception Index (Figure 2). A country's score indicates the perceived level of public-sector corruption on a scale from 0 (highly corrupt) to 100 (very clean). Some factors are overlooked, e.g. in Singapore, there is no minimum wage for migrant workers. The index for 2015 showed that:

- Denmark was the least politically corrupt country.
- 68% of countries have serious corruption problems.
- Somalia was the most politically corrupt country.

Political corruption in Lebanon

- Lebanon is 123rd in the Corruption Perception Index for 2015.
- Public money has been lost due to corrupt state institutions and public services.
- 43% of companies pay bribes to government officials.
- The government has failed to hold a general election and has stopped providing basic services.
- Protests have sometimes been met with excessive force, raising questions about human rights.
- There are also wider concerns about human rights, use of torture and treatment of Syrian refugees and migrant workers.

A

Rank	Corruptions Perception Index score (2015)	Country
1	91	Denmark
2	90	Finland
3	89	Sweden
4	88	New Zealand
5=	87	Netherlands
5=	87	Norway
7	86	Switzerland
8	85	Singapore
9	83	Canada
10=	81	Germany
10=	81	Luxembourg
10=	81	UK

B

Rank	Corruptions Perception Index score (2015)	Country
158=	17	Haiti
158=	17	Guinea-Bissau
158=	17	Venezuela
161=	16	Iraq
161=	16	Libya
163=	15	Angola
163=	15	South Sudan
165	12	Sudan
166	11	Afghanistan
167=	8	North Korea
167=	8	Somalia

 Figure 2 *The top (**A**) and bottom (**B**) countries on the Corruption Perception Index for 2015*

Ten-second summary

- Some countries are advocates for and promote human rights (e.g. the USA), whilst others prioritise economic development (e.g. Singapore).
- Although China and India's economies have both grown rapidly, levels of democratic freedom and human rights differ due to different political systems.
- All countries have some degree of corruption, which impacts on human rights.

 Over to you

Summarise the content of this section in:

a three sentences
b three short notes
c three words
d one word.

You need to know:

- that, in Australia, differences in rights between indigenous and non-indigenous peoples have led to discrimination and inequality
- that laws to protect the rights of indigenous peoples and to promote anti-racism have made variable progress
- that gender inequality is a problem in Australia.

Big idea

There are variations in human rights within countries.

Free speech versus human rights in Australia

The Australian Human Rights Commission states that it is unlawful for a person to offend someone else because of their race or ethnicity.

- In 2016, a controversial cartoon in an Australian newspaper depicted a stereotypical indigenous Aboriginal man. It led to allegations of racial hatred and debate about the right of free speech versus racial discrimination.
- Aboriginal and Torres Strait Islanders (ATSI) have long suffered from discrimination, lower living standards and poor quality of life.

ATSI peoples in Australia

In comparison to non-indigenous Australians, ATSI peoples:

- have an average life expectancy that is 10 years lower (70 years instead of 80)
- have higher levels of drug, alcohol abuse and homelessness, and are more likely to smoke
- see their traditional way of life disappearing, including their beliefs, customs and knowledge
- are more likely to live in remote areas (43.8% compared to 2.2%) (Figure 1)
- are imprisoned at a rate that is 15 times higher
- are several times more likely to be unemployed
- lack basic literacy skills. (In 2015, 30% of ATSI adults lacked basic literacy skills. However, by 2011, 54% of those aged 20–24 had achieved the Australian equivalent of A level qualifications).

◀ **Figure 1** *Nearly half of ATSI people live in remote areas of Australia, which are often semi-arid*

ATSI peoples and the law

Before 1967, ATSI peoples were:

- not considered to be Australian citizens
- not allowed to vote
- forced to live on reservations
- often had their children forcibly removed to be raised in white-run institutions, resulting in '**stolen generations**' of ATSI peoples.

Since a referendum in 1967 leading to citizenship rights for ATSI peoples, the Australian Human Rights Commission protects the rights of ATSI peoples in two ways:

Preventing racial discrimination

- The Racial Discrimination Act makes racial hatred an offence.
- It protects indigenous peoples against discrimination in employment, education and accommodation.
- It ensures that they can access services and public places.

Advocating social justice

- Australia recognises the distinctive rights of indigenous citizens.
- It maintains their right to a culture and identity.
- It recognises the rights of ATSI peoples to self-determination and land.

Health and education issues for ATSI peoples

Past differences in the rights of ATSI peoples, compared with those of non-indigenous Australians, are being reflected in the current differences in their health, education and employment opportunities. There are noticeable differences in causes of death and ATSI peoples have much lower life expectancy (Figure 2).

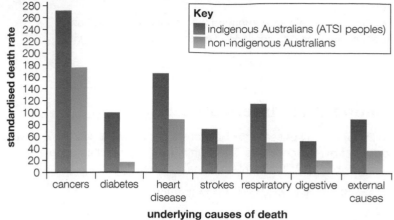

Quality of life

Many ATSI peoples:

- may not have enough money for day-to-day necessities
- lack access to a decent education. 20% of school-age children are not enrolled in school and,
  of those who are, many do not attend regularly.
  Only 78% achieved the required level of literacy at age 8
- lack good (or any) job opportunities
- lack connections to family and friends
- lack safety in their communities.

 Figure 2 *Selected underlying causes of death for indigenous and non-indigenous Australians in 2011*

'Closing the Gap'

A 2009 initiative, 'Closing the Gap' aims to improve quality of life (see also Section 4.2) and has made variable progress.

✓ Between 2005 and 2012, the life expectancy gap between ATSI and non-indigenous peoples closed by 0.8 years for men.
✓ Between 1998 and 2012, ATSI death rates from circulatory diseases fell by 45%.
✓ Between 1998 and 2012, the infant mortality rate for ATSI infants fell by 64%.

But:

✗ The life expectancy gap only closed by 0.1 years for women.
✗ Between 2008 and 2012, death rates from all avoidable causes were still three times higher for ATSI peoples than for the Australian population as a whole.
✗ ATSI Deaths from respiratory diseases fell by only 27%.

Equal rights for all Australians

The Australian government is committed to improving the quality of life for indigenous peoples. The main components of a 2016 motion to promote anti-racism were:

1 the entitlement of equal rights for all
2 a non-discriminatory immigration policy
3 reconciliation with ATSI peoples
4 Australia as a culturally diverse society
5 denouncing racial intolerance.

Gender inequality in Australia

Non-indigenous Australian women won the right to vote 69 years before ATSI women. Many women believe gender inequality in Australia is still an issue. They are concerned about:

- widespread discrimination and lack of safety
- everyday sexism in the workplace
- pay equality falling between 2015 and 2016
- poor support for working mothers.

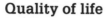
Ten-second summary

- Discrimination and unequal rights have led to lower living standards and quality of life for ATSI peoples in Australia.
- The Australian government's initiatives have made variable progress but it remains committed to promoting equal rights.
- Gender inequality remains an issue in Australia.

Over to you

Create a mind map to show why there are differences in human rights within Australia.

You need to know:

- the four different forms of geopolitical intervention in defence of human rights
- that there is rarely agreement on the validity of intervention
- that intervention can challenge ideas of national sovereignty.

Big idea

There are different forms of geopolitical intervention in defence of human rights.

Geopolitical interventions

In 2005, the UN passed the 'Responsibility to Protect' (R2P) resolution, adopted by 150 countries. If a country fails to protect its people, that responsibility falls to the international community.

Interventions generally take one of four forms.

1 Development aid

- This financial aid, sometimes called official development assistance (ODA), is given to developing countries.
- Most comes from the governments of developed countries, **IGOs** and **NGOs**.
- Most ODA from the USA in 2013/14 went to Myanmar, Afghanistan and India.
- Aid may go directly from one country to another (**bilateral aid**) or through an IGO (**multilateral aid**).
- Levels of ODA differ between countries (Figure 1) and over time (Figure 2).
- In 1970, the UN adopted a voluntary target for countries to commit 0.7% of their GNI each year to ODA. Most fail to meet this, although in 2015 the UK government made the target a legal requirement.
- The total amount of ODA has increased since 1960 and since 2000 there has been an increase of 82.5%. However, when measured as a percentage of GNI, the level of ODA has decreased.

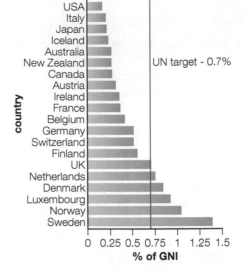

⊙ **Figure 1** *The level of ODA from donor countries as a percentage of GNI in 2015*

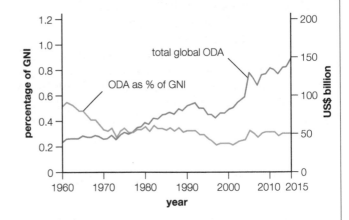

❯ **Figure 2** *Total global ODA, in US$ billion and as a percentage of GNI, 1960–2015*

2 Trade embargoes

- A trade embargo is a ban that restricts all trade, or trade in certain items (e.g. military supplies), with a country.
- It encourages a country to change its actions.
- Embargoes are often used in response to threats to international security or human rights abuses.
- In 2017, the UK had arms embargoes on 17 countries.
- The UN Security Council imposed an arms embargo on Libya in response to human rights abuses.

3 Military aid

- This consists of money, weapons or expertise given to developing countries to help them protect borders and fight terrorism or given to rebel groups fighting authoritarian governments.
- Since 2016, countries can include it in their 0.7% ODA target.
- Charities worry that less money is being spent on poverty.
- The USA is the largest contributor, usually to protect US interests.

4 Military action

This is generally considered to be a last resort, and can be of two types:

- **Direct action**, e.g air strikes and ground troops. In 2003, the USA and UK took direct action against the government of Saddam Hussein in Iraq.
- **Indirect action**, e.g. providing military assistance. In 2017, the UK took indirect action by training Nigerian forces to improve security.

Military action can be by agreement with the country or taken to protect people from their own government.

The validity of intervention

The international community's 'R2P' is not always straightforward. Sometimes decisions are supported, but there can be disagreement about whether intervention is justified. Disagreement occurs when:

- there are different views on the organisations or countries involved
- the intervention could be for the intervening country's self-interest
- there is a risk that intervention will make matters worse
- there is concern over a disregard for **national sovereignty**.

National sovereignty

National sovereignty is the idea that each nation has the right to govern without interference.

- However, this idea has limits, especially in matters concerning abuse and the rights of a nation's own people.
- There is tension between the ideas of national sovereignty and 'R2P', as became obvious in Libya from 2011.

 Figure 3 *Libya's geopolitical location in North Africa*

See page 187 of the student book for more information on the situation in Libya.

Factfile: tensions between national sovereignty and 'R2P' in Libya

Location: Libya, North Africa
Capital – Tripoli
World's tenth largest oil reserves

Human rights violations – In 2011, demonstrators against Colonel Gaddafi's government were brutally repressed with hundreds killed. The government failed to meet its responsibility to protect its citizens.

Invention – UN authorised bombing raids by UK and French air forces in support of civilians and rebels. Arms embargoes and air strikes supported rebels against government forces.

Concerns – Intervention was not widely supported. Five countries on the UN Security Council did not vote. They were concerned about insufficient evidence, that the real reason for intervention was regime change and about inconsistencies in interventions in the name of human rights.

Success – Gaddafi was dead by October 2011. Libya is still extremely unstable. Militias are in conflict with each other and the new government.

Ten-second summary

- Geopolitical interventions usually take one of four forms: development aid, trade embargoes, military aid or military action.
- The validity of intervention is rarely agreed on.
- Intervention can challenge the idea of national sovereignty, e.g. in Libya.

 Over to you

List the arguments in favour of, and against, intervention in the affairs of another country. Create a mnemonic to help remember them.

You need to know:

- that development aid can come in the form of loans from IGOs or charitable gifts from NGOs and national governments
- that aid has positive and negative impacts
- how economic development can have negative environmental impacts and disregard indigenous peoples' human rights.

Big idea

Development aid has a range of different impacts.

Types of development aid

Aid comes in different forms and can be provided for specific projects or for wider development aims.

Charitable gifts

- These are funded by public **donations** through charities or government funding.
- They can be either bi-lateral (involving two organisations, including government aid) or multi-lateral (involving many organisations, including NGOs). Between 2008 and 2011, 60% of ODA was bi-lateral.
- Governments favour bi-lateral aid as they can control its spending.
- Multi-lateral aid can be more 'legitimate', as NGOs do not have political self-interest.
- For example, US$13.5 billion was donated to Haiti after the 2010 earthquake: 75% from countries and 25% from charities.

Loans

- These are provided by IGOs, e.g. the World Bank and IMF.
- Between 2011 and 2015, US$150 billion was loaned by the World Bank to reduce poverty and stimulate economic growth.
- Loans can be used to build infrastructure, fight corruption and manage resources. For example, in 2015, the World Bank loaned US$40 million to Benin for infrastructure and flood defence schemes.
- Concerns over loans include the conditions that need to be met, the prioritising of economic development over environmental protection, and reductions in public spending on healthcare and education.

Development aid – a global success?

Aid can have both positive and negative impacts (Figure 1). Overall, minority groups are most at risk from the negative impacts, which can sometimes undermine human rights, development and democracy.

Positive impacts	Negative impacts
Aid related to public health, particularly vaccination programmes, has almost eradicated some diseases, e.g. polio.	Much aid is lost to corruption, especially as many countries with poor governance, such as dictatorships, receive aid.
In the fight against malaria, aid has provided mosquito nets, access to medicines and better diagnosis. Between 2000 and 2015, the new infection rate fell by 37% and mortality rates fell by 60%.	Aid can be used by political elites to ensure they remain in power, repress citizens and enrich themselves through corruption, e.g. Zambia's former president, Frederick Chiluba.
The UN's 'Decade for Women' has moved gender equality up the aid priority list. By 2014, over US$30 million was being targeted at gender equality programmes. Maternal mortality rates have fallen by 44% since 1990. More girls now attend school, but this is not universal.	A significant proportion of some developing countries' income comes from ODA (Figure 2). They can become aid dependent and unable to function without it. It is easier for governments to rely on aid than to improve their industries or increase tax. Food prices can fall, countries cannot make long-term plans and are at risk if the aid stops.

Country	ODA received as a % of GNI
Afghanistan	23.3
Central African Republic	35.9
Liberia	44.3
Micronesia	33.9
Tuvalu	63.3

⬆ **Figure 2** *The level of overseas aid received as a percentage of GNI in 2014*

⬆ **Figure 1** *Positive and negative impacts of aid*

Development aid versus environmental protection

Sometimes, development projects run by TNCs can result in the loss of farmland, ecosystems and traditional livelihoods, as well as social and health problems for local communities. The discovery of large oil reserves in Nigeria's Niger Delta and their exploitation by TNCs is an example of this.

Living in the Niger Delta

- The Niger Delta contains various ecological zones, e.g. mangroves, swamps and rainforest (Figure 3).
- It has a population of 31 million and 40 ethnic groups.
- 70% of people live below the poverty line.
- Only 20% of the area is accessible by road.
- Healthcare and education are underfunded.
- Access to clean water is limited due to pollution and poor sanitation.
- Poor care is taken over the environment and safety.

Impacts of oil in the Niger Delta

Economic development, driven by exploiting oil reserves, has had serious environmental and social impacts in the Niger Delta. The money brought by the oil industry is not reaching local communities.

- 75% of the Nigerian government's income comes from oil.
- There is weak governance and reliance on the US$10 billion from oil.
- Oil companies can operate unchallenged.
- Oil spills are common – 550 in 2014, polluting water and soil, and destroying mangroves, rainforests and local fish stocks (Figure 4).
- Burning off of natural gas during oil extraction causes acid rain. This damages farmland and releases pollutants into the air, causing breathing problems and higher risk of cancers.
- Clean up is at a cost of US$1 billion.

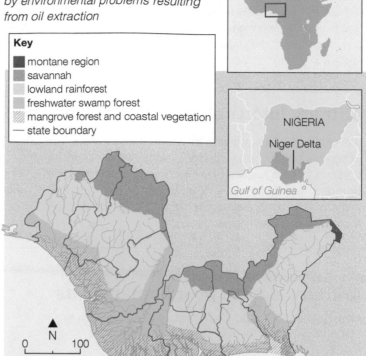

Figure 3 *The ecological zones of the Niger Delta are greatly affected by environmental problems resulting from oil extraction*

Key
- montane region
- savannah
- lowland rainforest
- freshwater swamp forest
- mangrove forest and coastal vegetation
- — state boundary

NIGERIA
Niger Delta
Gulf of Guinea

N
0   100
km

 Figure 4 *In Rukkpokwu, a fire caused by an oil spill burned for six weeks, polluting water supplies and destroying forests and farmland*

Ten-second summary

- Development aid can come in the form of loans or charitable gifts.
- Aid has both positive and negative impacts.
- Economic development can sometimes have negative environmental and social impacts, e.g. oil extraction in the Niger Delta.

Over to you

Draw a mind map to show different types of aid, their advantages and disadvantages, and examples of each.

You need to know:

- that military intervention is often justified in terms of human rights
- that there are arguments for and against giving military aid to countries with poor human rights records
- that military intervention as part of the 'war on terror' is justified by some on grounds of human rights, but criticised for use of torture.

Big idea

Military aid is often justified on the grounds of protecting human rights.

Military interventions – justifiable or not?

It is very difficult to justify military intervention in the affairs of another country. Internationally, such intervention must be seen as both justified and proportionate, or it could be branded illegitimate and a threat to international security. It is easier to justify an intervention on the grounds of protecting of human rights.

In Libya in 2011 (see Section 4.7), military intervention was justified by human rights abuses. However, the desire for regime change and to secure Libya's oil exports could also have been underlying reasons.

Military aid and poor human rights

Military aid may still be given to countries with poor human rights records (e.g. Colombia, Figure 1).

The UK has a list of 30 countries that it considers a 'human rights priority'. Yet since 2014 it has trained security or military personnel in, or from, 17 of those countries, including Afghanistan, Myanmar and Iraq. There are two opposing views on military aid.

Countries that send military aid argue that:

- to stop sending military aid would threaten national interests or global security
- stopping the aid would do nothing to stop human rights abuses
- they can pressure recipient nations with conditions attached to the aid.

Critics argue that:

- strategic alliances and trade are more important than human rights to the governments sending military aid
- ignoring human rights abuses condones them
- supporting a repressive government is wrong
- the aid could be used to commit further abuses.

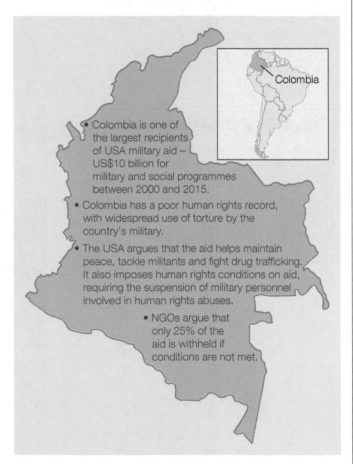

- Colombia is one of the largest recipients of USA military aid – US$10 billion for military and social programmes between 2000 and 2015.
- Colombia has a poor human rights record, with widespread use of torture by the country's military.
- The USA argues that the aid helps maintain peace, tackle militants and fight drug trafficking. It also imposes human rights conditions on aid, requiring the suspension of military personnel involved in human rights abuses.
- NGOs argue that only 25% of the aid is withheld if conditions are not met.

⌃ **Figure 1** *Colombia has a poor human rights record, but the USA still sends military aid*

The war on terror

After 9/11, US President George W. Bush declared a 'war on terror' to protect the USA and its allies from future attacks by terrorist organisations.

- President Bush justified the decision to send troops to Iraq and Afghanistan, claiming that these countries supported terrorists (a view that was contested).
- Protecting human rights was also used as justification. For example, in 2003, the USA and UK explained to the UN how 'weapons of mass destruction', claimed to be in Iraq, could threaten global security. Again, this was contested.

The use of torture and threats to human rights

Under international law and the UDHR, torture is illegal.

- The USA is criticised for using '**extraordinary rendition**', transferring terror suspects to other countries and interrogating them in secrecy. Other countries (e.g. the UK) are accused of helping the USA.
- The USA is criticised for using torture in interrogation, but it argues that its methods are 'enhanced interrogation techniques', which provide information to prevent future terrorist attacks.

This has damaged the USA's global reputation.

- Al-Qaeda have used it as an argument to recruit new members.
- Captured Americans are more likely to be treated harshly in return.
- Trust has been lost between the USA and community-based organisations.
- Countries that assisted the USA have also been condemned by human rights groups and the UN because of their support.

The USA and Guantanamo Bay

The USA established a military base at Guantanamo Bay in Cuba in 1903. Since the war in Afghanistan, it has been used to hold detainees in the 'war on terror'. The US has been accused of using torture there (Figure 2).

▼ **Figure 2** *The USA's holding and treatment of prisoners at Guantanamo Bay is controversial*

1903
The USA establishes Guantanamo Bay in Cuba.

2001
After 9/11, the base holds detainees in the 'war on terror'. It is not covered by US laws and so detainees are denied legal protection and held indefinitely.

2004
The International Red Cross finds evidence of torture at the camp. The USA defends itself by arguing that it uses legal 'enhanced interrogation techniques' which are necessary for gaining information to prevent future attacks.

2009
President Obama bans 'non-coercive' methods of interrogation and orders the camp to close.

2014
The USA admits to the use of torture.

2017
The camp still holds 41 detainees.

Ten-second summary

- Military intervention is sometimes justified on the grounds of protecting human rights.
- Military aid is sometimes given to countries with poor human rights records.
- The USA justifies its 'war on terror' in terms of human rights, but is criticised for using torture and 'extraordinary rendition'.

Over to you

List the arguments why **a** military aid and **b** torture are often rejected by many as inappropriate ways of intervening in another country's affairs.

You need to know:

- how the success of geopolitical interventions can be measured
- that democracy can be a measure of success and encouraged by democracy aid
- how success can be measured in terms of economic growth.

Big idea

There are several criteria for measuring the success of geopolitical interventions.

How successful is geopolitical intervention?

Most improvements in global development have resulted from **geopolitical interventions** and global policies, e.g. the MDGs.

- As interventions are often politically sensitive and expensive, they need to be shown to work.
- This demonstrates accountability to voters and whether actions need improvement.
- Intervention can be assessed using 'hard' data (i.e. statistics, as in Figure 1) and 'soft' data (e.g. degrees of freedom of speech or gender equality).

Measuring the success of intervention

This can be difficult because:

- 'Success' has varying definitions.
- Some countries cannot collect data accurately.
- Interventions span many years (e.g. the MDGs) and changing circumstances.
- External factors (e.g. global food prices) can affect outcomes.
- Data outcomes are interpreted differently by different people.

Indicator	Uganda		Bangladesh	
	2000	**2015**	**2000**	**2015**
Infant mortality (per 1000 live births)	86	38	61	31
Maternal mortality (deaths per 100 000 live births)	620	343	399	176
Average life expectancy	46	58	65	72
% of literacy among women	50.2 (1995)	71.5	26.1 (1995)	58.5
HDI score	0.393	0.483 (2014)	0.468	0.570 (2014)

- Health indicators include life expectancy and infant mortality
- Education indicators include literacy and length of schooling
- Wealth, including GDP per capita, is one aspect of the HDI index

⬆ **Figure 1** *'Hard' indicators for health, education and wealth/HDI show improvements in Uganda and Bangladesh between 2000 and 2015*

Democracy as a measure of success

For many countries and IGOs, the promotion of democracy is a key goal of intervention. Democracy is crucial for economic growth and stability within a country because:

- It leads to other economic and social changes.

- Countries become less willing to support criminal organisations.
- It is easier for military and economic ties to be forged.
- Countries are less likely to go to war.

Democracy aid and freedom of expression are therefore important factors in promoting democracy.

Democracy aid

Democracy aid is used to promote democracy.

- Funds for building democracy come from Western governments.
- Democracy aid totalled US$10 billion in 2015.
- It supports fair elections and the development of political parties.
- It strengthens governmental institutions such as parliaments.
- It defends civil and political rights.
- 26 out of 57 countries that became democracies between 1980 and 1995 received democracy aid from the USA.

Freedom of expression

Freedom of expression is an important part of promoting democracy.

- It is a fundamental right, outlined in the UDHR.
- It guarantees the right to speak and write openly.
- It protects against injustice.
- It enables criticism of government and leaders.
- It is believed to be a cornerstone of democracy.
- For example, the 'Strengthening Freedom of Expression Protection' project in the Gambia works to improve relations between the media and government.

Economic growth as a measure of success

For many countries, economic growth is the key goal of overseas aid and the most important means of reducing poverty. One estimate suggests that a 10% increase in a country's average income reduces poverty by 20–30%. It brings about improved infrastructure, health, education and environmental protection. Examples of success include:

- South Korea and Singapore, which once received US aid and are now beneficial trading partners with the USA.
- China, whose overseas aid to sub-Saharan Africa has been based on economic development. This has resulted in higher employment and economic growth, as well as ensuring benefits for China.

Aid for trade

Trade is essential to economic growth. 'Aid for trade' is a World Trade Organisation (WTO) initiative to help developing countries increase their trade (Figure 2). It helps them to:

- develop better trade strategies
- negotiate better trade deals
- build infrastructure such as roads and communications.

Between 2000 and 2013, 48% of Uganda's ODA was 'Aid for trade'. Aid was used mainly in the transportation, energy and primary employment sectors (Figure 3). In the same period, Uganda's:

- exported goods increased by 144%
- GDP per capita doubled
- poverty levels dropped by 10%
- average life expectancy rose from 46 to 58
- infant mortality fell from 86 per 1000 live births to 38
- HDI rose from 0.393 to 0.483.

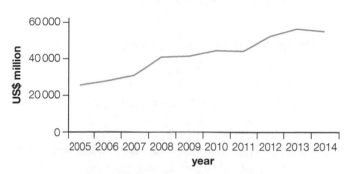

Figure 2 The trend for 'Aid for trade' is growing. Its share of global ODA rose from 26% in 2006/07 to 35% in 2010

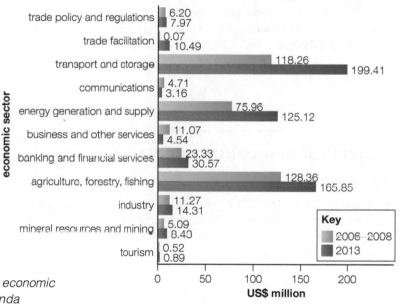

Figure 3 The distribution by economic section of 'aid for trade' in Uganda

 Over to you

List:

a the different ways in which geopolitical intervention may take place
b the ways in which success can be measured.

You need to know:

- that development aid can have varying levels of success
- how aid can increase and decrease inequality within a country
- that donor countries can use development aid as an extension of foreign policy.

Big idea

Development aid has a mixed record of success.

Development aid – successful?

Development aid aims to fix a wide range of problems, so success can vary.
For example, aid in West Africa and in Haiti has had different results.

Ebola, West Africa – a success story!

- The first cases of Ebola were recorded in March 2014.
- In 21 months, 28 616 cases were reported, with 11 310 deaths.
- The outbreak was not easily contained. Symptoms take up to 21 days to appear, so those infected could travel widely and spread the virus.
- The WHO declared it an emergency and sent teams of health workers.
- The UN Security Council called a rare emergency session to assess the implications.
- Longer-term development aid funded general health services.
- In January 2016, the WHO declared the region disease-free.

Haiti – a failed effort?

- Haiti is vulnerable to natural hazards, has high levels of poverty and relies on aid. Its aid dependency has limited its progress.
- It is one of the world's poorest and worst-governed countries.
- Jobs are done by aid workers rather than local people, so government systems are weak.
- Local skilled people work for NGOs rather than Haitian organisations.
- Aid money is spent on US contracts, rather than local ones.
- Haiti is one of the most unequal countries, as shown by its **Gini coefficient**.
- Economic development has been small.
- Human rights are also a problem.

Impacts of inequality on development

High inequality (Figure 1) in a country has several impacts.

- Health indicators (e.g. life expectancy and infant mortality) are poor.
- The poorest people cannot afford healthcare.
- Improvements in living standards are hindered as people are less economically productive.

⊙ **Figure 1** *Levels of inequality around the world based on Gini coefficient*

The UK and Haiti have a big difference in Gini.

Developing countries have wider inequality.

The darker the shading, the more unequal the country.

All countries have some degree of inequality.

Key
least unequal — most unequal

Can aid lead to increased inequality?

Sometimes, inequality can actually increase in countries receiving large amounts of aid.

For example, Bangladesh receives a large amount of aid (US$2.4 billion in 2014), but its Gini score has gone up (Figure 2). This is could be because:

- There is corruption by the political and economic elite (see Section 4.8).
- Donor countries can act in their own interest and decide where and how to spend aid.
- Aid agencies can favour large projects in return for publicity, rather than give to smaller, more effective projects.

However, Latin American countries receive large amounts of aid, and income inequality decreased between 2000 and 2010 in 16 out of 17 countries.

- The Gini coefficient fell by 0.94% on average each year.
- Health and education indicators also improved.

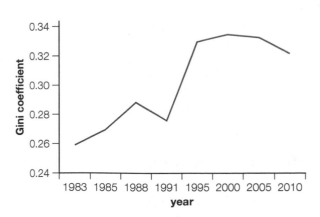

Figure 2 *The Gini coefficient for Bangladesh, 1983–2010*

Development aid – an extension of foreign policy?

Donor nations often use aid as an extension of their foreign policies. If the goals of their development aid are to promote their foreign policy, they might measure success of the aid programmes on the benefits they receive in return.

There tend to be three motives:

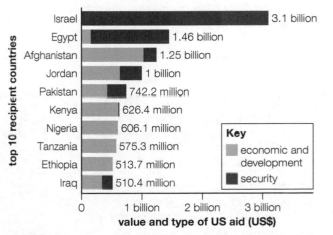

Figure 3 *The top ten recipients of US aid in 2017*

To develop military alliances

- Jordan received over US$750 million between 2014 and 2016 in return for military support against so-called Islamic state (Figure 3).
- It carried out air strikes and allowed overseas forces to use its military bases.

To access resources

- China has provided aid for the development of resources and infrastructure (e.g. a rail link between inland ports in Zambia and Tanzania) in sub-Saharan Africa.
- Its need for raw materials has influenced this aid and investment policy.

To gain political support in IGOs and NGOs

- India wants a permanent seat on the UN Security Council. Its FDI programmes have be targeted at countries that might back it.
- Similarly, Germany helped with the development of HEP plants in Nigeria in return for support for its permanent membership.

Ten-second summary

- The level of success of development aid can vary.
- Development aid can both increase and decrease economic inequality within countries.
- Development aid can be an extension of foreign policy and its success depends on the benefits that donor countries receive.

Over to you

Create a table to compare the positive and negative aspects of development aid.

You need to know:

- that direct and indirect military intervention can have some short-term gains and long-term costs
- how non-military intervention can sometimes be more effective
- that a lack of action can have negative impacts on human rights, development and the environment.

Big idea

Direct and indirect military interventions have a mixed record of success.

Military intervention

Military intervention is never an easy option. There are always economic, political, environmental and social costs.

Direct intervention

- Involves sending troops and equipment.
- Leads to loss of lives on both sides, as well as physical and mental injuries.
- Costs a lot of money, e.g. the 2003 Iraq War cost US$2 trillion.
- Can also have a political price, e.g. UK Prime Minister Tony Blair lost his reputation for involving the UK in the Iraq War.

Indirect intervention

- Involves providing economic or military assistance.
- Is a preferable option for governments.
- Involves much lower costs and risks.

Direct intervention – the costs of the Iraq War

In 2003, the USA and its allies (including the UK) invaded Iraq in order to:

- remove dictator Saddam Hussein from power
- protect civilians against Saddam's rumoured chemical and biological 'weapons of mass destruction' (WMD)
- prevent many human rights abuses.

There is debate over the success of the military intervention. In the short-term:

✓ Saddam Hussein and his oppressive security forces were removed.
✓ Early development efforts succeeded.
✓ A US-funded vaccination programme reduced infant mortality by 75%.
✓ Iraq held its first free election for over 50 years in 2005.

△ **Figure 1** *Iraqi civilians fleeing Mosul in 2017 as Iraqi government forces fight to drive out IS militants, who captured the city in 2014*

But in the long-term there were problems.

✗ No traces of WMD were found.
✗ Iraq was left without systems to restore security and democracy, protect human rights and grow economically.
✗ Islamist militant groups (such as Al-Qaeda and IS) took advantage of instability and established themselves in Iraq. They continue to fight and kill civilians.
✗ Animosity between Iraq's Sunni and Shia groups has worsened.
✗ Its political instability has allowed corruption to grow.
✗ Human rights remain a problem. In 2015, Human Rights Watch found evidence of attacks on civilians by government forces, unlawful detentions and a lack of freedom of expression.

The success of non-military intervention

Non-military intervention can sometimes lead to long-term improvements in human rights and development. For example, Timor-Leste has gained independence and ended violence and human rights abuses (Figure 2).

▶ **Figure 2** *Timor-Leste is an example of a successful non-military intervention*

Timor-Leste in South-East Asia was a Portuguese colony and declared independence in 1975. 9 days later, it was invaded by Indonesia. By 1999, 25% of the population had been killed by violence, disease and famine. There were many human rights abuses, such as torture.

The UN took control to set up structures to maintain law and order and promote development. In 2002, Timor-Leste became independent, although human rights are still a concern.

In 1982, the UN tried to resolve the conflict and organised a vote of independence, with 78.5% of votes in favour. But, anti-independence militia, supported by Indonesia, killed 7000 and displaced 40 000 people.

To pressurise Indonesia, there were arms embargoes from the UK and the USA and diplomatic efforts from the UN. Finally, Indonesia withdrew. This left a lack of skilled people to run the country.

Map labels: Banda Sea, Dili, TIMOR-LESTE (EAST TIMOR), INDONESIA, Timor Sea, N, 0 100 km

The effects of no military action

Zimbabwe is an example where a lack of action by the UN and international community has had global consequences (Figure 3).

About Zimbabwe	• It has a history of human rights abuses against its citizens. • Opponents of the government and Robert Mugabe (President from 1980 to 2017) were attacked and imprisoned. • It suffers from great poverty.
Reasons for the lack of intervention	• As a former British colony, Western nations are sensitive to intervention linked to colonialism. • Neighbouring African nations argued that Mugabe was not a threat and Western nations would not take action without their support. • It is unlikely that the UN Security Council would agree to intervention. A 2008 arms embargo failed after Russia and China voted against it.
Social results	• 72% of Zimbabweans live in poverty, with rural poverty increasing because seasonal plantation work has low wages. • Average life expectancy is low – 59 for men and 62 for women. • There is much corruption within the government. • Human rights abuses are common, especially violence against political opponents.
Environmental results	• Deforestation rates are increasing. Large areas of forest were lost between 1990 and 2010. • Trees are cut down by the rural poor for firewood and to be used by tobacco farmers. • Tobacco farming is very important as it accounts for 25% of Zimbabwe's imports and provides jobs for the poor. • Laws passed to prevent damage are hard to enforce.

▲ **Figure 3** *The effects in Zimbabwe of lack of military action by the UN and international community*

Ten-second summary

- Military intervention may have short-term gains but often has more serious long-term costs, e.g. in the Iraq War.
- Non-military intervention can be more effective and leads to improvements in human rights and development.
- Sometimes, when military action is not taken, it can lead to negative economic, social and environmental impacts, e.g. in Zimbabwe.

Over to you

Create a mind map to show:

a the costs of military intervention (in red)
b the benefits of military intervention (in green).

Chapter 5
Migration, identity and sovereignty

What do you have to know?

This chapter studies ways in which tensions can result between globalisation and the traditional definitions of national sovereignty and territorial integrity. International migration changes ethnic composition of populations but also changes attitudes to national identity. Nationalist movements challenge models of economic change and redefine ideas of national identity. Global governance has a mixed record of success in dealing with issues arising from these tensions.

The specification is framed around four enquiry questions:

1 What are the impacts of globalisation on international migration?
2 How are nation states defined and how have they evolved in a globalising world?
3 What are the impacts of global organisations on managing global issues and conflicts?
4 What are the threats to national sovereignty in a more globalised world?

The table below should help you.

- Get to know the key ideas. They are important because 20-mark questions will be based on these.
- Copy the table and complete the key words and phrases by looking at Topic 8B in the specification. Section 8B.1 has been done for you.

Key idea	Key words and phrases you need to know
8B.1 Globalisation has led to an increase in migration both within countries and among them.	globalisation, rural-urban migration, international migration and migration policies, source areas, destinations, economic migrants, refugees, asylum seekers
8B.2 The causes of migration are varied, complex and subject to change.	
8B.3 The consequences of international migration are varied and disputed.	
8B.4 Nation states are highly varied and have very different histories.	
8B.5 Nationalism has played a role in the development of the modern world.	
8B.6 Globalisation has led to the deregulation of capital markets and the emergence of new state forms.	
8B.7 Global organisations are not new but have been important in the post-1945 world.	
8B.8 IGOs established after the Second World War have controlled the rules of world trade and financial flows.	
8B.9 IGOs have been formed to manage the environmental problems facing the world, with varying success.	
8B.10 National identity is an elusive and contested concept.	
8B.11 There are challenges to national identity.	
8B.12 There are consequences of disunity within nations.	

You need to know:

- how professional football reflects migration on a wider scale.

Big idea

Football reflects trends in the global economy towards freer flows of labour, capital and power.

A football league of nations?

The English Premier League has changed since it was launched in 1992 (Figure 1).

These changes have been revolutionary, and the revolution has been repeated across Europe (Figure 2).

- The main cause of this revolution was the 1995 Bosman Ruling, which enabled free movement of European footballers anywhere across the EU. Note the big rise after the 1995–96 season.
- At the same time, migration laws permitted entry of highly skilled individuals from outside the EU with the offer of a job contract.

Does this mark the end of English football? In reality, a global league has emerged, rather than an English league. Football simply reflects what is happening in the global economy, where there are freer flows of labour, capital and power.

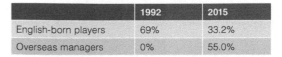

	1992	2015
English-born players	69%	33.2%
Overseas managers	0%	55.0%

⬆ **Figure 1** *Changes in the Premier League since it was launched in 1992*

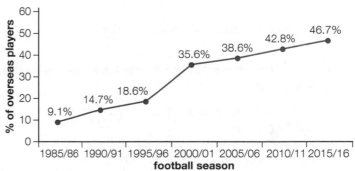

⬆ **Figure 2** *The rising percentage of overseas players in the big five European football leagues (English, German, Spanish, Italian, French) between the 1985/86 and 2015/16 seasons*

Globalisation and football

Global media coverage, sponsorship by major TNCs and very high salaries all make the world's top football leagues attractive to world-class players, coaches and managers.

Like all migrations, there are problems and benefits for the source countries.

- The movement of footballers, sometimes called the **muscle drain**, is **de-skilling** African clubs of their most talented players.
- However, many overseas players send a proportion of their income home as **remittance payments**.

A world without borders?

IT and digital communications systems are now spreading ideas and information faster than ever, helping to create a more connected world. However, there are wider implications of this change to a 'borderless' world.

- Globalisation can change the cultural and ethnic composition of some communities and even nations.
- National identities are affected.
- Tensions and conflicts can rise as people adapt to a new sense of national identity.

 Ten-second summary

- The English Premier League has seen a rise in the number of overseas players, managers and owners.
- Football reflects wider-scale migration and the freer flows of labour, capital and power in the global economy.
- Migration has impacts on source countries and on the 'borderless world'.

Over to you

Take the letters of the word 'FOOTBALL' and use them to create a mnemonic of the key geographical information on this page.

You need to know:

- how globalisation has encouraged rural-urban and international migration
- how migration is affected by policies and the global economy
- how events alter the pattern of international migration.

Big idea

Globalisation has led to an increase in migration both within countries and between them.

Changing patterns of demand for labour

Globalisation works on the principle of free flows of investment capital. Two processes therefore affect the demand for labour:

- At a **national** scale, people move from traditional rural economies to work in industrial cities.
- At an **international** scale, there is easier movement of people, especially for those with skills.

Rural-urban migration in China

China's rapid industrialisation has been accompanied by rapid urbanisation fuelled by internal rural-urban migration. There are two main flows:

- of rural migrants usually to small cities within the rural interior
- of migrants from smaller cities to major east-coast cities and industrial areas, e.g. Shanghai.

In 1980, over 80% of Chinese people lived in rural areas. By 2012, 51% of the population was urban.

- However, there are barriers to migration within China known as the **hukou** system. Those moving to cities from rural areas must be 'registered' and buy a permit, which is expensive.

The EU Schengen Agreement

The Schengen Agreement abolished many internal border controls within the EU.

- It enabled passport-free movement across (at the time) 22 EU member states (excluding the UK) and 4 non-EU member states.
- Free movement across the Schengen area has helped to fill job vacancies in EU countries. However, some people claim that it allows migrant labour to be paid lower wages and free movement of criminals and terrorists.
- In 2016, six Schengen countries, including Germany and France, re-introduced internal border controls.

See page 214 of the student book for further information about the hukou system.

Trends in international migration

International migration is movement across national boundaries. Around 4% of the global population live outside their country of birth. Figure 1 shows global flows of migrants in 2014.

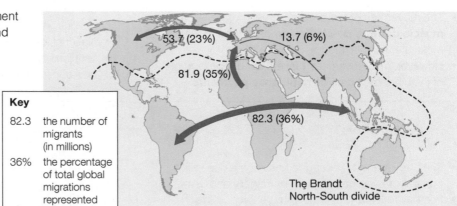

Key

82.3	the number of migrants (in millions)
36%	the percentage of total global migrations represented

53.7 (23%)   13.7 (6%)

81.9 (35%)

82.3 (36%)

The Brandt North-South divide

▲ **Figure 1** *Flows of migrants between and within the global 'North' and 'South' in 2014*

Variations in international migration

The proportion of international migrants within each country varies, depending on each country's attitudes and policies towards immigration and engagement with the global economy.

Japan

There is a closed-door policy to immigration and little evidence of change in the political mind-set to alter this.

- Japan's population is falling and 27% are aged over 65. The UN suggests that Japan needs 17 million immigrants by 2050 to maintain its population at 2007 levels.

Australia

There is a pro-migration policy.

- 70% of Australia's immigrants are accepted based on skills shortages.
- Australia's immigrants contribute on average 10% more per capita to Australia's GDP each year than non-immigrants.
- Immigrants also offset the numbers due to retire.

Changes in international migration

Most migrants fall into three categories:

- **Voluntary economic migrants** – moving for work.
- **Refugees** – forced to leave their country because of war, natural disaster or persecution.
- **Asylum seekers** – fleeing to another country to apply for the right to international protection.

Once migrants are established in their host countries, a fourth group may then follow them – **family members**.

	2004–05		2009–10		2014–15
Poland	62.6	India	75.3	Romania	169.8
India	32.7	Poland	69.9	Poland	128.4
Pakistan	20.3	Lithuania	23.4	Italy	64.4
South Africa	19.3	Latvia	23.2	Spain	58.6
Australia	16.6	Pakistan	23.0	Bulgaria	44.1
Lithuania	15.6	Bangladesh	21.2	India	39.6
France	13.3	Romania	17.7	Portugal	37.5
China	12.6	France	16.5	France	31.0
Portugal	12.2	Nigeria	16.2	Hungary	26.5
Slovakia	10.5	Nepal	14.6	Lithuania	25.0
Total	**439.8**		**572.8**		**917.4**

Figure 2 *Numbers of overseas nationals allocated a National Insurance number on entry to the UK, from 2005 to 2015 (numbers in thousands)*

The UK

The UK has two main sources of international migrants:

- **The Commonwealth** – Afro-Caribbean migrants arrived from 1948 to fill labour shortages, followed by those from India and Pakistan.
- **The EU** – economic migration to the UK has been high, which has helped to fill labour shortages (Figure 2). After the UK's decision to leave the EU, there are questions about future migration.

Migration elsewhere

- **Europe** – from 2014 to 2017, there were huge increases in migrants from North Africa and the Middle East, especially Syria.
- **Middle East** – many labourers moved from India, Pakistan and Bangladesh to Qatar to help build infrastructure for the 2022 FIFA World Cup.
- **Globally** – natural hazards have also forced people to move. Haiti's 2010 earthquake displaced 300 000 Haitians.

Ten-second summary

- Rural-urban migration and international migration occur to fill job vacancies.
- Japan and Australia have very different immigration policies.
- Factors such as labour demands and war affect the pattern of international migration.

Over to you

List all the migrations in this section to show where migrants are travelling from and to. Explain the main reason for each migration.

You need to know:

- the causes of migration
- how freedom of movement poses challenges for national identity and sovereignty
- how movement of labour is unrestricted within many nations but not at a global level.

Big idea

The causes of migration are varied, complex and subject to change.

Crossing the Mediterranean

See page 218 of the student book for migrants' accounts of travelling to Europe.

In 2015, just over one million migrants left North Africa and the Middle East for Europe.

- Many of those leaving the Middle East and Africa were **refugees** fleeing from conflict and poverty.
- On arrival in Europe, their aim was often to claim **asylum** (apply for the legal right to protection in their destination country).

EU migration to the UK

Although many migrants are refugees, most are economic migrants moving for **work**.

- Most of those coming to the UK from the EU in 2015–16 gave work as their main reason (Figure 1).
- Nearly half of migrants from other parts of the world came to study.
- 60% of EU nationals arrived from Eastern Europe after eight countries there joined the EU in 2004. They helped to fill gaps in the UK labour market.
- The high numbers arriving reflected unemployment figures and low wages in Eastern Europe.

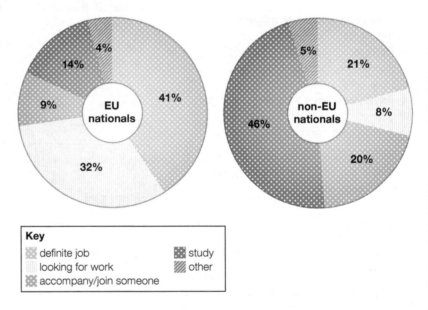

Key

- definite job
- looking for work
- accompany/join someone
- study
- other

🔺 **Figure 1** *Reasons given for immigration to the UK up to March 2016 (not including British nationals returning to the UK)*

Migration, national identity and sovereignty

The theory of globalisation is based on a belief in freer flows of people, capital and trade, known as **neo-liberalism**. It involves:

- **trade** liberalisation by removing subsidies, tariffs, quotas and trade restrictions
- the **freedom to invest** anywhere or transfer capital, known as deregulation of financial markets
- **open borders**.

But these freedoms mean that national borders become unimportant, which raises challenges. For example:

- What effect does the free movement of people have on the ways in which people identify with a country (i.e. their **national identity**)?
- What does being a part of a trading bloc mean for a country's '**sovereignty**' (i.e. its ability to protect its independent rule of law or governance)?

National identity

National identity refers to a national feeling of being a cohesive whole, which sometimes rises above other identities (e.g. Scottish nationalism).

- But common beliefs or values can vary over time, in intensity or by age group.
- Migration also affects identity, as different cultures enter a country.

Sovereignty

Sovereignty means the authority of a state to govern.

- However, globalisation can reduce its impact. For example, the UK voted to leave the EU in 2016.
- Among the reasons given by 'Leavers' was loss of sovereignty resulting from EU membership.
- However, few laws actually directly originated from Europe.

Open borders and immigration

Immigration is controversial and can cause resentment within host populations, which may sense threats to national identity.

- Extreme political parties are becoming increasingly significant in Europe.
- Since 2014, huge numbers of Syrian refugees and economic migrants have caused tensions between Greece, other Balkan countries (the entry points to Northern Europe) and Turkey.

Internal migration within the UK

In any year, about 10% of people move within the UK (Figure 2).

- This **regional movement** of people is unrestricted and is often linked to the changing labour market.
- For example, de-industrialisation in northern Britain since the 1980s has driven many people south in search of employment.
- The regeneration of large cities (e.g. Manchester) has led to in-migration of younger people for work and the lifestyle. By contrast, many older adults with families move from cities into rural areas.

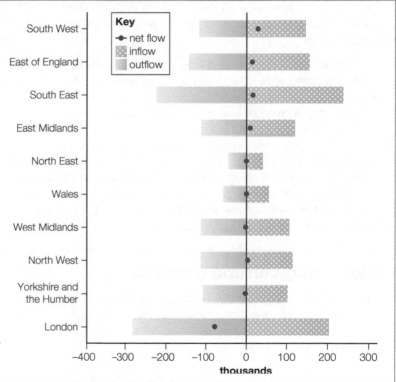

▶ **Figure 2** *Internal migration flows for the nine English regions, plus Wales, for the year up to June 2015*

The social consequences of migration

- High internal migration can lead to rising house prices in destination cities or areas, strained healthcare provision and falling wages.
- Conversely, large out-migrations from source regions can leave those areas with skills shortages and an ageing population.

Globally, movements of labour *are* restricted, but that doesn't stop migration.

 Ten-second summary

- Most migrants move for work, but others are refugees.
- Migration can cause host populations to feel that national identity and sovereignty are being threatened.
- Internal migrations occur due to the changing labour market.

 Over to you

From memory, define the following terms:

a refugee   d trade liberalisation
b asylum seeker   e national identity
c economic migrant f sovereignty.

You need to know:

- how the rate of assimilation of migrants varies
- why migration causes political tensions
- reasons why there are variations in people's ability to migrate across national borders.

Big idea

The consequences of international migration are varied and disputed.

All change!

Migration can change the **cultural** and **ethnic** composition of countries. The degree of change depends on the rate of **assimilation** of migrants into the host nation.

- The extent to which migrant groups are assimilated or remain segregated varies within and between countries.
- In South Africa, for example, nearly three decades of equal rights since the abolition of **apartheid** have hardly altered its **ethnic segregation**.

Even in Western countries, ethnic groups may still be segregated by residence (Figure 1). Unlike in South Africa, this ethnic segregation is usually due to economic and cultural factors.

- Cheaper rental properties in inner cities have traditionally attracted migrants to settle close to their workplaces.
- Over time, ethnic **enclaves** become permanent, such as Hindu and Sikh communities in Southall, West London.
- West Indian immigrants in the 1950s were recruited to drive London's buses, so many settled near London's bus garages, e.g. Brixton.
- Once established, cultural factors – such as the growth of specialist shops and places of worship – help to maintain these separate ethnic enclaves.
- Gradually, many ethnic communities have integrated into British economic life while at the same time retaining their cultural distinctiveness.

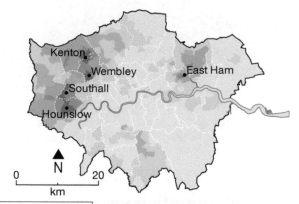

Key
% Asian/Asian British: Indian
- 37+
- 26–37
- 16–26
- 9–16
- 4–9
- 0–4

🔺 **Figure 1** *The clustering into particular London enclaves of Asian/Asian British people of Indian descent, from the 2011 Census*

Measuring cultural diversity

Cultural diversity varies within and between nations. Figure 2 shows global diversity on a map of **cultural fractionisation**. This uses an index to measure people's attitudes and determine how diverse countries are. The index varies between 1 (total diversity) and 0 (no diversity).

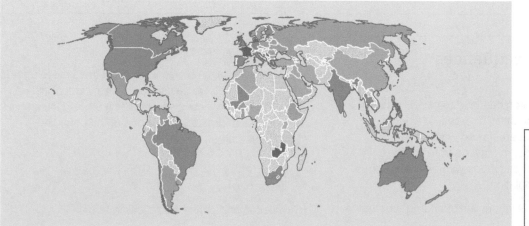

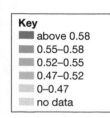

Key
- above 0.58
- 0.55–0.58
- 0.52–0.55
- 0.47–0.52
- 0–0.47
- no data

🔺 **Figure 2** *Cultural fractionisation or diversity; darker-shaded countries are more diverse than those shaded lighter*

Migration and political tension

Japan

Japan's economy is sluggish and its population is decreasing and ageing.

- This means there are fewer workers and increased spending on healthcare.
- For many countries, immigration would be the answer. However, Japan has a deep-rooted cultural aversion to this solution: only 1.63% of the Japanese population is immigrant.
- The Japanese hold the belief that they are a 'homogenous' people.
- They fear that immigrants may disrupt the 'harmony and co-operation' that characterise Japanese society.

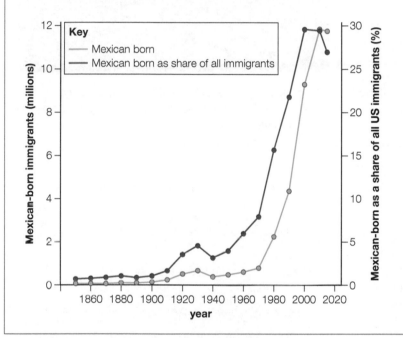

Mexico and USA

Large-scale migration from Mexico to the USA increased with a surge (from 1970 onwards) of both legal and illegal Mexican immigrants (Figure 3).

- By 2015, there were over 11.7 million Mexican immigrants in the USA.
- In a survey, about half of the US people asked (51%) said that immigrants strengthened the country, while 41% felt that they were a burden.
- The views expressed were largely split along political lines, with Republican voters much more anti-immigration (63%).
- Often, during his 2016 election campaign, Donald Trump called for a wall to be built along the entire US–Mexican border to prevent further illegal immigration.
- However, many Americans actually believe that the main focus should be on developing a plan to manage Mexican immigrants already living in the USA illegally, rather than trying to halt the declining flow of illegal immigrants still arriving.

 Figure 3 *Mexican immigrant population in the USA, 1850–2015*

Ability and opportunities

A new life in another country is easier to achieve for some than others. For example, it can depend on a person's **skills**.

- Migrants to Australia generally enter as skilled workers, but they need a minimum of 65 points on Australia's point-based system.
- Points are awarded according to age, qualifications and competence in English. Preference is also given to those with an existing job offer.

In other cases, the ease of migration can depend on **existing wealth**.

- Many illegal Mexican migrants pay people smugglers between US$4000 and US$10 000 to cross the border into the USA.

Ease of migration can also depend on practical **opportunities**, including:

- the presence or absence of international border controls
- the presence of established and settled family members in the destination country.

 Ten-second summary

- Ethnic groups can remain segregated within countries.
- Opinions vary on migrants and migration within and between countries.
- The ability to migrate can depend on skills, wealth and practical opportunities.

Over to you

Create a table to show the causes and consequences of migration.

You need to know:

- that nation states vary in their ethnic, cultural and linguistic (language) unity
- how national borders have been created
- how contested borders can lead to conflict and migration.

Big idea

Nation states are highly varied and have very different histories.

Iceland

Iceland's national characteristics result from its:

- geographical location and isolation
- landscape (Figure 1)
- dependence on the sea.

By present standards, the Icelandic population is **monocultural**. Iceland's laws and society fiercely protect its cultural heritage and national identity.

- All children's names must come from an approved list.
- The Icelandic language has remained unchanged since AD 870s.
- There is religious homogeneity. 74% of Icelanders belong to the Evangelical Lutheran Church of Iceland.
- Those born overseas constituted only 8.9% of the population in 2015.

▲ **Figure 1** *Most of Iceland is completely empty and unsettled, with over half the population living in or close to Reykjavik, the capital*

Singapore

This small country has a vibrant mix of languages, culture, religions, festivals and food.

- Singapore was first established in 1819 as a British colonial trading post. Its subsequent growth was largely due to immigration from China, India and Malaysia.
- Its population today reflects its globalised present and its multicultural past: 74% are Chinese, 13% Malay and 9% Indian, with others of European descent or ex-patriots working overseas.

The concept of nation states

The concept of a nation, or sovereign state, is one in which the population is united by factors such as:

- language
- ethnic and cultural background
- customs.

These help to create a sense of national identity.

- However, nation states develop and change over time and can vary with population migrations.
- Some nations do not exist physically but their identity exists, e.g. Kurdish territories. Dispersed populations like these are known as **diasporas**.

National borders

Borders separate nations and are the result of different factors.

- **Natural borders** consist of physical features such as rivers, lakes, or mountains.
- **Colonial history and political intervention** – in 1884–85, at a conference in Berlin, the continent of Africa was divided up along **geometric boundaries**. The countries became colonies of the major European powers.
- The new country borders took no account of existing tribal or linguistic boundaries. African peoples were given no say.

Rwanda

Belgium took over colonial rule of Rwanda (Figure 2) after the First World War.

- The Belgians favoured the Tutsi minority (14% of the population), who had privileges over the Hutu majority (85%).
- When Rwanda gained independence in 1962, its government was contested.
- During the 1970s and 1980s, Hutus were given preferential jobs in the public services and military.

In April 1994, the president of Rwanda was killed when his plane was shot down.

- That event led to the **genocide** (mass killing) of an estimated 800 000 Tutsis and moderate Hutus.
- In July 1994, the Hutu government fled to Zaire (now the Democratic Republic of Congo), together with 2 million Hutu refugees, and an interim government of national unity was established.

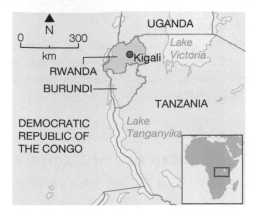

▲ *Figure 2* Rwanda

Contested borders

Ukraine and Crimea

58% of Crimea's population are ethnic Russians, 24% are ethnic Ukrainians, 12% are Muslim Tatars. The country has a varied history.

- It was part of Russia for much of the period from 1783 to 1954.
- In 1954, it was transferred to Ukraine. Both Ukraine and Russia were Soviet republics in the former USSR.
- After 1991, when the USSR broke up, Ukraine became independent. But part of Russia's navy was based in Crimea, at Sevastopol (Figure 3).
- In 2010, Ukraine agreed to extend Russia's lease on Sevastopol until 2042 (in exchange for cheaper Russian gas).

In 2014, Ukraine's pro-Russian president was driven from power.

- The new government favoured membership of the EU and NATO. Russian-backed forces seized control of Crimea.
- Its Russian-speaking majority voted to join Russia in a highly controversial referendum. 850 000 Ukrainians fled Crimea as a result.

▲ *Figure 3* Crimea and Ukraine

Taiwan and China

Taiwan has effectively been an independent state since 1950, but China still claims sovereignty.

- China insists that other nations should not have official formal relations with both China and Taiwan.
- Despite its diplomatic isolation, Taiwan is one of Asia's economic success stories.

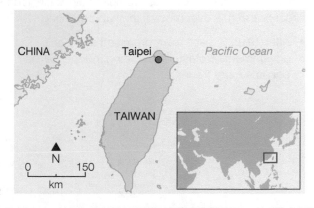

▶ *Figure 4* Taiwan and China

Ten-second summary

- Iceland is monocultural; Singapore has a mix of cultures.
- Rwanda's borders are a result of its colonial history, which took no account of different ethnic groups.
- Borders between Ukraine and Russia are contested, and Taiwan is not universally recognised.

Over to you

Annotate a map of the world to show five key facts about each of the case studies in this section. Put it up on your wall as a reminder and read it every two days.

You need to know:

- that 19th-century nationalism was important in the development of empires
- that the emergence of new nation states has caused conflicts
- how patterns of migration between former colonies and the 'parent' country are still evident.

Big idea

Nationalism has played a role in the development of the modern world.

The emergence of Europe's nation states

In Europe before the 16th century, many people viewed themselves as part of a local community rather than a nation. But by the end of the 19th century, most of the present European states had emerged in one form or another.

Nationalism in the 19th century

Nationalism is based on peoples' identification with a nation, in the belief that they share a common identity, language, history and customs that bind them together.

- In **France,** the French Revolution helped to establish nationalism as a force by giving power to ordinary citizens. Loyalty to France as a nation grew and a new national identity emerged. Napoleon Bonaparte, the Emperor of France, believed in expansionism.
- Many saw French nationalism as a threat. It led to rising nationalism in countries like **Austria** and **Russia**, which sought revenge for French aggression.

- Growing German industrial wealth and infrastructure encouraged a growing sense of being German. The new state of Germany was created in 1871.
- After the First World War (1918), national boundaries shifted again. Ethnic Germans were now scattered in foreign territories. In the period leading up to the Second World War, Hitler re-incorporated most of these territories into Germany.

Empires and their consequences

Nationalism grew beyond Europe in the 19th and early 20th centuries. Most European nations extended their overseas interests and colonies (Figure 1).

India

Trade with, and political influence over, a large part of India by the British East India Company eventually led to direct British rule over most of the sub-continent.

- After 1918, Indians were promised some self-government. Protests demanding independence became more frequent.

- This resulted in actions such as the Amritsar Massacre of 1919, when British troops opened fire on unarmed Indian protestors, which encouraged Indian nationalism.
- India was offered complete independence in 1946.

Elsewhere

- After 1945, colonial rulers of sub-Saharan Africa gradually gave up political control to new independent governments.
- When British Prime Minister Harold Macmillan gave his famous 'wind of change' speech in 1960, it was clear that new states were about to emerge.

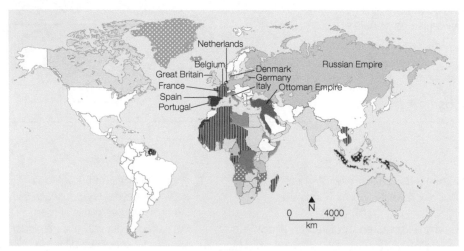

◀ Figure 1 *The extent of European empires by 1900*

Key
- Belgium
- France
- Germany
- Great Britain
- Italy
- Netherlands
- Portugal
- Denmark
- Spain

The costs of disintegrating empires

Vietnam – North or South

Vietnam (Figure 2) is a former French colony.

- In 1954, nationalist leader Ho Chi Minh 'reclaimed' Vietnam from France.
- The USA was concerned about the spread of communism in South-East Asia. To reach an agreement, Vietnam was divided into North and South.
- Vietnamese nationalists, supported by communist China, controlled the North. In the South, independent non-communist rule began, supported by US troops.
- A war ensued. 1–4 million Vietnamese were killed.
- The South was finally defeated in 1975 and an independent, united Vietnam emerged.

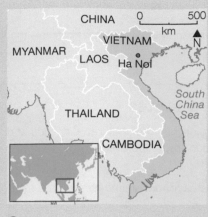

▲ **Figure 2** *Vietnam*

South Sudan – a new African country

When the major European nations agreed Africa's borders (see Section 5.5), Sudan was divided into northern and southern territories.

- Britain and Egypt modernised the mainly Arab north, leaving the mainly black African south to tribal communities.
- The north prospered more, so people in the south felt marginalised and left out of Sudan's development.
- In 2011, the country of South Sudan was created (Figure 3), but the presence of 60 different ethnic groups has made central government control difficult.
- Between 2013 and 2017, there was civil war, with over 2 million people displaced by the conflict.

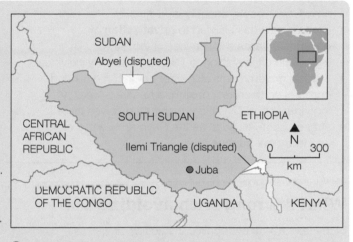

▲ **Figure 3** *South Sudan*

Migrations and colonial ties

After the Second World War, employers in the UK recruited workers from Commonwealth countries, such as Jamaica, supported by the British government.

- The purpose was to fill urgent job vacancies in areas such as transport (e.g. for London Underground and buses) and the new NHS.
- These migrants were usually well qualified.
- Migrants from India and Pakistan soon followed those from the Caribbean. Again, many were well qualified, including doctors.
- Other less well-qualified migrants were attracted by opportunities in manufacturing industries. The textile towns of Lancashire and Yorkshire attracted many men from rural Pakistan.

As the level of international migration increased, the ethnic composition changed and a **cultural mosaic** of people evolved across the country. The UK's relative homogeneity gave way to a more heterogeneous mix of peoples.

Colonial legacies

There is still a strong correlation between former territories of past colonial rulers and the languages spoken there.

- For example, although there are over 40 ethnic groups in Uganda, English remains the national language, which helps with Ugandan trade and development.

Ten-second summary

- Nationalism has been a source of conflict in Europe and beyond.
- The emergence of new nation states has caused conflicts.
- Migration from former colonies has changed the ethnic composition of host countries.

Over to you

Write ten questions and answers about the migration topic so far. Test yourself in a few days' time.

You need to know:

- how globalisation has encouraged the growth of tax havens
- that tax havens are largely accepted
- how global inequalities can drive alternative economic models.

Big idea

Globalisation has led to the deregulation of capital markets and the emergence of new types of state.

New rule – new states

Globalisation has revolutionised global trade, finance and the movement of labour. Trade liberalisation has lifted many trade regulations.

- There has also been a reduction in governmental roles in the economy.
- **Deregulation** in the 1970s and 1980s meant that state interference was reduced or removed.
- Post-deregulation, capital could be transferred anywhere – freely, cheaply, and quickly.
- Globalisation has also led to **privatisation** of government assets in services or industries, with ownership often shifting to TNCs and wealthy individuals.
- Several governments also offer low income tax and corporation tax rates, designed to attract wealthy individuals and TNCs to register themselves there.
- These countries, such as the Cayman Islands, are known as **tax havens**.

The Cayman Islands

The Cayman Islands are among the world's largest **offshore** financial centres.

- They are only permitted to work with businesses resident outside their territory.
- 40 of the world's top investment banks and insurance companies are licensed there. In 2014, the islands held US$1.5 trillion in assets.
- The Cayman Islands have a 0% personal income tax rate and low corporation tax rates, which makes them irresistible for individuals and companies seeking to pay low or no tax.

What's wrong with avoiding tax?

Many tax havens offer political stability, as well as secure banking and legal systems, without breaking any laws. However, several are associated with political and economic instability and corruption.

Major TNCs and wealthy **expatriates** (those who live in a country where they are not citizens) 'rest their cash' in safer tax havens.

- However, doing so is controversial. For example, in 2015, Apple held 89% of its total cash abroad (Figure 1).
- By doing so, it avoided paying 35% in corporate taxes, which would otherwise have been due in the countries where its profits were being made.

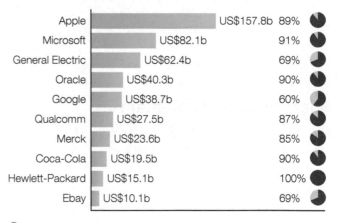

Apple	US$157.8b	89%
Microsoft	US$82.1b	91%
General Electric	US$62.4b	69%
Oracle	US$40.3b	90%
Google	US$38.7b	60%
Qualcomm	US$27.5b	87%
Merck	US$23.6b	85%
Coca-Cola	US$19.5b	90%
Hewlett-Packard	US$15.1b	100%
Ebay	US$10.1b	69%

⚫ **Figure 1** *The percentage of total wealth held abroad by top US companies in March 2015*

Attitudes towards tax havens

Most governments and IGOs accept the growth of tax havens.

- National governments seek investment from TNCs to generate employment and wealth, and they have the freedom to set their own tax rates. However, many NGOs have raised objections.
- In 2015, the EU declared that the ability of TNCs to move revenues earned throughout the EU to low-tax member states like Ireland was unfair.

Growing inequalities

Despite increased wealth since the 1980s, growth in global income has not been evenly distributed (Figure 2).

- Most growth occurred in China as it became the world's workshop.
- Outsourcing and offshoring reduced employment and incomes in the USA and Europe.
- The top 1% – the '**global elite**' – have gained hugely.

Geographers such as Danny Dorling argue that economic and social stability are threatened by increasing inequality. Every step down from the richest to the poorest means:

- reduced life expectancy, education and work prospects
- increased mental health problems.

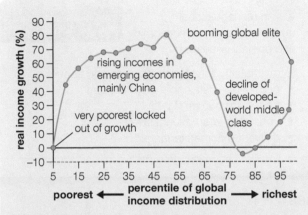

... the era of globalisation seemed to offer little for people in-between: households in the 75th to 85th percentile of income distribution (who were poorer than the top 15% but richer than everyone else) seemed scarcely better off in 2008 than in 1988. They constituted a decile of discontent, squeezed between their own countries' plutocrats and Asia's middle class. This dramatic dip in the chart seemed to explain a lot. Cue Donald Trump. Cue nationalism. Cue Brexit ...

Adapted from The Economist, *September 2016*

▲ *Figure 2* *Growing inequalities: 'the elephant graph'; percentiles of income growth between 1988 and 2008*

Alternatives to globalisation – Bolivia

Not all countries support the processes of globalisation. By 2010, eight South American countries, including Bolivia, had elected left-wing governments. Bolivia established its National Coalition for Change (CONALCAM) in 2007. Its policies are:

- nationalised resources (e.g. oil) – the profits went to the government rather than to private shareholders
- reduced primary exports and boosted domestic manufacturing of previously imported products (known as **import substitution**)
- redistributed wealth to the *campesino* (peasant classes) by guaranteeing prices for food products.

The state is now the largest player in the economy. Bolivians began benefiting from:

- increased gas connections (by 835%), electricity (by 150%) and telecommunications (by 300%)
- improved healthcare, education enrolment, pensions and incomes
- reduced wealth inequalities (Figure 3) and lower government debt.

The demand for domestically manufactured goods drove sustained annual growth of over 5% between 2006 and 2012. Bolivia is now one of the world's fastest growing economies.

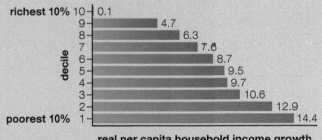

▲ *Figure 3* *Bolivia under President Morales begins to close its wealth gap*

- Some TNCs and wealthy expatriates take advantage of low tax rates in tax havens.
- Some NGOs have raised objections to tax havens.
- Bolivia has promoted an alternative economic model.

 Over to you

Create a flow diagram to show how globalisation has encouraged the growth of tax havens.

You need to know:

- about the role of the United Nations in global governance
- how interventions by the UN have a mixed record of success
- how some member states have operated independently of the UN.

Big idea

Global organisations are not new but have been increasingly important in the post-1945 world.

The UN – the first IGO

In October 1945, the UN became the world's first true Inter-Governmental Organisation (IGO), with 50 members.

- Its structure and roles were agreed by the USA, the UK, the USSR and China.
- As allies, they aimed to maintain global peace and security after the Second World War.
- The USA, the UK, the USSR, China and France became the five permanent members of the UN's Security Council.

The role of the UN

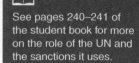
See pages 240–241 of the student book for more on the role of the UN and the sanctions it uses.

The UN's work has evolved to help manage global environmental, socio-economic and political issues, e.g.

- maintaining international peace and security
- promoting sustainable development
- protecting human rights
- upholding international law
- delivering humanitarian aid.

Personal influences and national disputes

The UN's role is influenced by the UN Secretary General. For example, in 2006, Secretary General Ban Ki-Moon made climate change a UN priority.

Sometimes national disputes spill over into UN policy making. The Syrian Conflict has raised ideological differences between Russia (backing President Assad) and the USA (backing rebel groups). This kind of conflict challenges the UN's role as peacekeeper.

The UN's use of sanctions

The Security Council meets to respond to threats. Sometimes that involves introducing **economic sanctions**, **embargos** (bans) on selling certain goods or even using **direct military intervention** to protect people. Possible sanctions include:

- arms embargos
- more general trade embargos
- restrictions on loans
- freezing financial or property assets
- travel restrictions (e.g. on diplomats).

Bosnia – direct military involvement

In 1993, there were allegations of **ethnic cleansing** by Bosnian Serb forces against Bosnian Muslims.

- The UN designated a safe zone in Srebrenica in north-eastern Bosnia.
- In 1995, Bosnian Serb forces captured Srebrenica, massacring 8000 Muslim men and boys.
- Several Dutch peacekeepers were taken hostage and threatened with execution if the Dutch interfered.
- Again, this raised questions about UN effectiveness.

Iran – economic sanctions

A suspicion that Iran was attempting to build nuclear weapons led to imposition of economic sanctions by the UN.

- At the time, Iran was the world's fourth largest oil-exporting country.
- The trade embargo started in the middle of 2012 and had an effect on Iranian oil exports (Figure 1).
- However, although the UN was shown as willing to act, Iran's annual GDP fell by only 5%. This raised questions about whether sanctions work

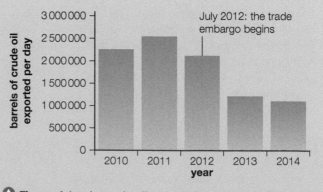

▲ *Figure 1* Iran's crude oil exports between 2010 and 2014

Taking unilateral action – the USA

Sometimes UN members take part in **unilateral** action, where one country, or group of countries, acts against another without formal UN approval.

- An extreme example occurred after the 9/11 attack in 2001.
- As part of the 'War on terror', a coalition of forces led by the USA, invaded Iraq in 2003 and deposed Saddam Hussein's government.

Unilateral action against Russia

In 2014, Russian-backed forces seized control of the Crimea region of Ukraine.

- In protest, the EU, the USA, Australia, Canada and Norway all imposed sanctions on 23 leading Russian politicians.
- The USA also led moves towards **sectoral sanctions**, targeting key areas of the Russian economy such as energy and banking.

The UN General Assembly met in 2014, but not all UN members agreed to act further.

- Member states took no action when Russia used its veto.
- Sanctions were strengthened, but EU member states were reluctant to go too far because they depended on Russian gas and oil supplies.

The impacts were substantial – up to a point.

- US$70–90 billion left Russia, as wealthy Russian investors sought secure overseas banks, devaluing Russia's currency.

But Russia retaliated, banning imported food from the EU and USA (Figure 2). As a result:

- Russia became less dependent on oil and gas exports and diversified its economy.
- Russian farmers gained home markets because of restrictions on imported food.
- The EU kept importing Russian energy supplies, despite sanctions.
- Food exports from the EU and USA were hit, e.g. Dutch tomato and cucumber sales to Russia fell by 80%.

Proposed unilateral action – the UK

- In 2013, the UK Government sought a resolution of the UN Security Council to condemn the use of chemical weapons by Syrian government forces.
- This required the backing of the five permanent members – China, Russia, the UK, the USA and France – and the vote was unlikely to pass.
- In the event, the UK Parliament voted against interventions in Syria.

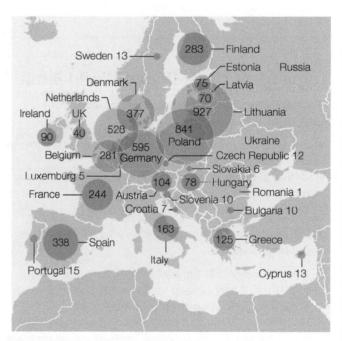

 Figure 2 *The lost value of EU food exports after Russia banned imported EU food; losses per country are in million euros*

 Ten-second summary

- The UN helps to manage global environmental, socio-economic and political issues.
- Sanctions in Iran and direct military involvement in Bosnia had varying degrees of success.
- Unilateral action has been taken against countries such as Iraq and Russia.

Over to you

Write down the three bullet points from the Ten-second summary. From memory, write three sentences to support each bullet point.

You need to know:

- about the importance of the IMF, World Bank and WTO in the global economy
- about impacts of Structural Adjustment and HIPC policies
- about the emergence of trading blocs.

Big idea

IGOs established after the Second World War have controlled the rules of world trade and financial flows.

International financial organisations

The **World Bank** and **International Monetary Fund (IMF)** were established in 1944 with the aim of stabilising global finances after the 1930s Great Depression and the costs of the Second World War.

- Both organisations follow the **Washington Consensus**, a belief that economic efficiency can only be achieved if regulations are removed.
- They follow the 'Western way' of organising capitalism.
- The **World Trade Organisation** (WTO) is part of the same family of organisations.

Financial IGOs

- **IMF** – aims to ensure global financial stability. Banks and member governments pay into a fund used for loans to help stabilise national currencies.
- **World Bank** – is funded in the same way. Its purpose is to finance loans for development.
- **WTO** – its main purpose is to promote global trade by reducing barriers, e.g. tariffs and duties.

See Section 3.5 for more information about these organisations.

Impacts of the IMF and World Bank

Between them, these IGOs promote **neo-liberalism**. Their objectives have been:

- in the 1950s, to support post-war reconstruction among *developed* countries
- in the 1970s and 1980s, to loan money for development projects in *developing* countries.

However, global interest rates rocketed in the 1980s, making loan repayments unaffordable for many developing countries.

- The IMF and World Bank would only help countries in difficulty if they agreed to conditions known as **Structural Adjustment Programmes (SAPs)**.
- Under these conditions, countries had to export more (to earn capital to repay loans) and reduce government spending.

SAPs also required debtor nations to accept other conditions:

- **expanding domestic markets**, allowing private companies to develop resources for export
- **reducing the role of government**, e.g. privatising state industries
- **removing restrictions on capital** for international investments
- **reducing government spending**, e.g. cuts to welfare
- **devaluing the currency** to make exports cheaper.

Critics argue that these conditions forced countries to sacrifice their **economic sovereignty**.

The Highly Indebted Poor Countries (HIPC) initiative

In 1996, the IMF and World Bank introduced the **HIPC**.

- It aimed to reduce national debts by partially writing them off in return for SAPs.
- The HIPC initiative affected 36 of the world's least-developed countries.

In 2005, UK Chancellor Gordon Brown steered the G8 (see Section 3.2) to cancel all debts owed to the World Bank and IMF by 18 HIPCs. There were conditions:

- Each country had to show good financial management and lack of corruption.
- National governments had to spend the savings gained on poverty reduction, education and healthcare.

Uganda and debt

In 1992, Uganda's debts totalled US$1.9 billion. In 2000, it was among the first to benefit from debt write-offs under the HIPC (Figure 1). The impacts were immediate.

- Government spending rose by 20%.
- Free primary schooling was introduced.
- Figure 1 shows other impacts.

	Before debt cancellation	Soon after cancellation	10–15 years after cancellation
GDP (US$ billion)	4.3 (1990)	7.9 (1992)	27.5 (2015)
% of people using improved water sources	44 (1990)	60.0 (2004)	79.0 (2015)
% of total population undernourished	24 (1990–92)	19.0 (2002–04)	25.5 (2015)
% of GDP spent on education	1.5 (1991)	5.2 (2002–05)	2.2 (2013)
% literacy rate age 15 and above	US$56.1 billion (1985–95)	66.8 (1995–2005)	78.4 (2015)
% of export income spend on debt repayment (debt service ratio)	1.4 (1990)	9.2 (2005)	1.8 (2016)

Figure 1 *The impacts of debt relief on Uganda*

Global and regional trade

Membership of major global trade and financial IGOs is almost universal. WTO members accounted for 96.4% of world trade in 2015.

- Regional trading blocs have also emerged. Groups of nations that operate without cross-border taxation (e.g. NAFTA), and which permit freedom of movement of goods, services and people (e.g. EU), are referred to as **single markets**.
- Such links can extend further (Figure 2).

Trading blocs – towards political union or not?

Centripetal forces (e.g. harmonisation of economic policies) draw member states together. This requires trust and economic consistency. However, some members may not want union.

- Nationalism can create **centrifugal** forces, driving organisations apart.
- The UK has always welcomed the economic benefits of the EU single market but some worried about diluted UK sovereignty.

Free trade area
- No internal trade barriers.
- Individual members retain their own currencies and economic policies.

Customs union
- No internal trade barriers.
- Common external tariffs.

Single market
- No internal trade barriers.
- Common external tariffs and free movement of labour.
- A common currency?

Political union
- Total unity – individual nations fuse as one; national boundaries disappear.
- Total freedom of movement of goods, services, labour and capital.
- Common economic and defence policies.

 Figure 2 *The development of trading blocs: from a free trade area to possible political union*

 Ten-second summary

- The IMF, World Bank and WTO aim to ensure global financial stability, finance global development and promote global trade respectively.
- SAPs involved countries meeting conditions in order to reorganise their loans.
- HIPCs involved some cancellation of debt.
- Regional trading blocs have emerged.

Over to you

Copy the 'You need to know' bullet points, then record everything you can remember for each one. Check your notes against this section.

You need to know:

- about issues concerning the quality of the global atmosphere and biosphere
- how IGOs have been involved in developing environmental laws
- about IGO management of the environment, including Antarctica.

Big idea

IGOs have been formed to manage the environmental problems facing the world, with varying success.

Managing threats to the biosphere

In 2016, the UN Environment Programme (UNEP) stated that wetlands were among the most diverse and productive ecosystems, yet also among the most threatened by urbanisation and economic development.

- The UNEP exists to promote sustainable development and manage the Earth's atmosphere and biosphere.

- Early attempts to conserve wetlands came in 1971, with the **Ramsar Convention**, an international treaty.
- Globally, over 2200 Ramsar sites are now managed by national governments, international NGOs (e.g. WWF) and local bodies like Natural England.
- Although successful elsewhere in the world, wetlands are in significant decline in the UK.

Managing threats to the atmosphere – the Montreal Protocol

The build-up of ozone-depleting substances (ODSs) – including chlorofluorocarbons (CFCs) – in the atmosphere increases the amount of harmful UV radiation reaching the Earth's surface from the sun.

- This radiation damages human health, ecosystems, bio-geochemical cycles and air quality.
- CFCs have already damaged the ozone layer sufficiently to cause the large 'hole' over Antarctica (Figure 1).

To address atmospheric deterioration, the Montreal Protocol on Substances that Deplete the Ozone Layer was signed in 1987.

- By 2010, virtually all countries had phased out ODSs (Figure 2).
- The global ozone layer should return to pre-1980 levels by 2050, and over Antarctica by 2070.

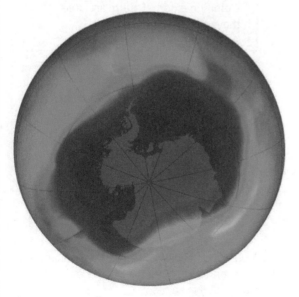

🔺 **Figure 1** *The largest recorded Antarctic ozone hole (in blue and purple), recorded by NASA on 24 September 2006*

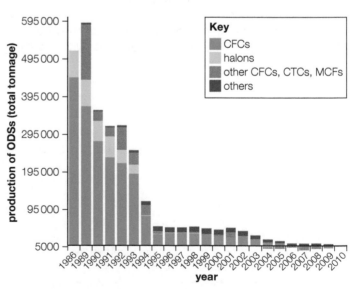

🔺 **Figure 2** *Reduction in the production of ozone-depleting substances since 1986*

Other international agreements by IGOs to protect the environment

Convention on International Trade in Endangered Species (CITES)

- CITES aims to ensure that international trade in wild animal and plant specimens does not threaten those species' survival.
- It protects over 35 000 species of flora and fauna.
- However, its enforcement has not been strict and therefore it has not been universally successful.

UN Convention on the Law of the Sea (UNCLOS)

In 1994, 157 countries signed UNCLOS.

- It provides guidelines for managing marine resources, creating Exclusive Economic Zones.
- Within these, coastal nations have sole exploitation rights over resources, including fishing and minerals.

The Water Convention

- The Water Convention aims to protect the quantity, quality and sustainable use of trans-boundary water resources (e.g. rivers and aquifers) by promoting co-operation between countries.
- It is also known as the Convention on Protection and Use of Trans-boundary Watercourses and International Lakes.
- 42 nations signed it in 1992.

Millennium Ecosystem Assessment (MEA)

The MEA began in 2001 to assess the consequences of ecosystem change, and the actions needed to conserve and use ecosystems sustainably.

- The MEA is needed. Half the world's natural habitats have been cleared for human use (mostly agriculture) and up to 1.5% is cleared annually.
- Current rates of extinction are 100 times faster than natural rates.

Protecting the Antarctic environment

Antarctica has no native population, government or laws.

- However, as scientific interest there – and potential for resource exploitation – increases, management of the region is needed.
- 12 countries signed an agreement in 1959 to create the Antarctic Treaty System. Another 41 have signed since.

This agreement sets strict rules:

- Antarctica is only to be used for scientific research.
- No military action or equipment is allowed.
- No territorial claims can be made on the region.
- Under the **Antarctic Protocol on Environmental Protection** (1991), no exploitation of any resources is allowed until at least 2041.

See pages 250–251 of the student book for source materials so you can evaluate the effectiveness of IGOs' actions to protect the environment.

Ten-second summary

- The UNEP manages threats to the atmosphere and biosphere.
- The Montreal Protocol manages threats to the atmosphere.
- A range of other actions by IGOs (e.g. CITES and UNCLOS) aim to protect the environment.
- The Antarctic Treaty System aims to protect Antarctica.

Over to you

For each example in this section, create a revision card that outlines:

- what the action is (e.g. Montreal Protocol)
- what it does
- how successful it has been, including at least two facts to support this.

Student Book
See pages 252–255

You need to know:
- how nationalism is reinforced
- how identity and loyalty are tied to a range of factors
- that questions of national identity and loyalty are complex.

Big idea
National identity is an elusive and contested concept.

Sport and nationalism

International sporting events are an opportunity to indulge in **nationalism** – a patriotic feeling.

- They can nurture pride, unity and loyalty – even if only briefly – and the effect is often strongest for the host nation.
- Identifying and believing in what your nation stands for often encourages loyalty and a sense of belonging.

Education and 'British values'

UK politicians like to speak about 'British values', but how can values that vary between people and that change over time be defined? One traditionalist newspaper identified a list of 'British' values (Figure 1).

Since 2014, English schools have been legally required to promote British values. These comprise beliefs and practice in:

- **democracy**, e.g. the right to vote
- the **rule of law**, e.g. the right to trial by jury
- **individual liberty**, e.g. the right to choose where you live
- **mutual respect and tolerance** of those with different faiths and those without faith.

- The rule of law – the same rules for everybody
- Parliament, the Monarch and the Supreme Court constitute supreme authority
- Tolerance – no one should be treated differently on the basis of belonging to a particular group
- Personal freedoms
- Freedom of speech
- A belief in private property, and the freedom to buy and sell
- Institutions that capture and reflect the British character, e.g. the monarchy, armed forces, Church of England, BBC
- British history and culture
- A love of sport and of fair play
- Patriotism

Figure 1 *Core 'British' values and characteristics suggested by* The Daily Telegraph *newspaper*

Politics and values

Politicians usually combine national duty with a political agenda.

- In 2005, there were protests across China and South Korea because the Japanese government approved school textbooks promoting nationalist views of Japan's history.
- The writers were accused of ignoring events like the 1937 Nanking Massacre (in which the Chinese government claims 400 000 Chinese civilians were killed by invading Japanese troops).
- Japan's Prime Minister claimed that their aim was 'to protect a country with a proud history'.

In the UK, the UK Independence Party (UKIP) secured the highest votes (27.9%) in the 2014 European Parliament elections and collaborated in the 2016 campaign to leave the EU.

- They support citizenship tests for migrants seeking permanent residency.
- These are supposedly an indicator of willingness to adopt the values and identity of the host country, by knowing something of its history and culture.
- However, by 2018, UKIP support was declining fast, suggesting some political influences are only temporary.

Distinctive identities

The opening and closing ceremonies of major international sporting events (e.g. Olympic Games in London, 2012) have become opportunities to showcase the host nation's perceived national identity. They provide an insight into the character of the host nation – not just by what is shown, but also in the way it is choreographed and produced.

Globalisation and identity

Identity with, and loyalty to, a home country varies between generations.

- Occasionally surges in nationalism occur, reflecting the mood of the times.

Politicians try to articulate national characteristics and values.

- France pursues a policy to protect its culture, and particularly its language, against influences such as Anglicisation.
- Japan is reluctant to increase immigration in order to protect its own culture.
- But globalisation – with fewer restrictions on movement, trade and investment, and global media – can alter people's views, loyalties and national identities.

Concepts such as tolerance, freedom, respect and fairness are not unique to Britain. They are as common within migrant communities as they are within British-born individuals and communities.

- Urban centres usually attract most migrants, and British cities have become centres of mixed populations with people speaking different languages.
- This shows how globalisation complicates definitions of national identity and loyalty.

London: the face of multi-cultural Britain

In 2016, London's population was 8.6 million.

- ONS data show that 3.1 million were born overseas and 44% was made up of black and other ethnic minorities (growing from 29% in 2001).
- Its wide ethnic mix makes the identity of both London and the rest of the UK complex.

In 2011, people's own perception of their national identity was included as a question in the 2011 UK Census for the first time (Figure 2).

- Ethnic minorities often identified more closely with 'Britishness' than their 'white British' counterparts.
- 'White British' people more frequently viewed themselves as 'English', 'Welsh', etc.
- By contrast, a high percentage of 'white other' people (perhaps European, perhaps American) felt little or no connection with 'Britishness', but instead retained their 'overseas' identity.

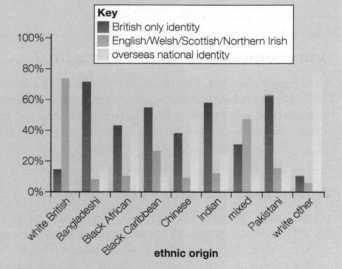

 Figure 2 *People's own perceived national identity, as expressed in the 2011 UK Census*

⏱ **Ten-second summary**

- Nationalism is reinforced through education, sport and politics.
- British values can be difficult to define but include ideas about the rule of law and liberty.
- The wide ethnic mix makes the identity of the UK complex.

✏ **Over to you**

Create a mind map to show how national identity and loyalty can be reinforced.

You need to know:

- that 'Made in Britain' is an increasingly complex idea
- how 'Westernisation' promotes a distinctive view of the benefits of capitalism
- how foreign ownership of property and businesses impacts on national identity.

Big idea

There are challenges to national identity.

Made in Britain?

In 2016, foreign TNCs spent over US$100 billion buying British companies. It's difficult to recognise goods 'Made in Britain' because they might:

- be made by British companies under overseas ownership
- consist of parts made abroad but assembled in the UK.

Figure 1 lists some British brands and their overseas owners. However, it works both ways, e.g. UK company Serco runs transport in Dubai.

Companies achieve economies of scale by merging, as well as by eliminating competitors. The UK government rarely intervenes in the sale of UK companies to overseas buyers, unless national security, financial stability or **media plurality** (the ownership of several forms of media by the same company) are at risk.

British brand	Foreign owner(s)
Scottish Power	Spain's Iberdrola
BAA airports operator	Spain's Ferrovial
Thames Water	German RWE
P&O	Dubai Ports
Abbey National bank	Spain's Santander
Cadbury's	USA Kraft/Modelez
Boots	Italian Equity firm
Harrods	Qatari Investment Bank
Fortnum & Mason	Canada's Weston
British Energy	French EDF
British Steel	India's TATA
National Lottery	Canadian Lottery

🔺 *Figure 1 Made in Britain?*

Does company nationality matter?

A company's nationality may not matter if it provides job security and boosts the economy. But take-overs mean that profits go abroad so corporate taxes are paid overseas. For example, the Italian owner of Boots the Chemist moved its headquarters from the UK to low-tax Switzerland in 2014. Its new tax bill was £9 million, instead of the £89 million it had paid one year earlier in the UK.

Westernisation and cultural values

In some ways, globalisation is another term for 'cultural imperialism'. Previous generations called it 'Americanisation', 'Westernisation' or 'modernisation'. Forbes (a US company) calculates the brand value of the top 500 companies. Of the top ten in 2015, eight were American.

Entertainment is increasingly provided by a small group of huge companies.

- Three companies (Sony BMG, Universal Music Group and Warner Music Group) own 80% of the global music market. All are American.
- In Australia, 70% of newspapers are owned by Rupert Murdoch's News Corporation.

Westernisation also affects **retailing**.

- The American model for retail is the mall. Most indoor centres (e.g. the Trafford Centre in Manchester) are built by large property companies with investment from pension funds and big banks. They change the nature and identity of retail in towns and cities.
- Within each mall, global brands tend to replace independent businesses.

A cultural takeover?

Disney owns foreign-language radio stations and TV channels, and international magazines. By 2015, it had opened over 140 learning centres in China equipped with Disney materials for teaching English to 150 000 children.

Disney promotes a distinctive view of the benefits of Western capitalism. It targets the middle classes in China and India, many of whom see Western brands as symbols of economic success. For many, Westernisation represents a bigger picture of social mobility and personal freedoms.

Property and land – ownership and identity

In 2013, UK estate agents Knight Frank placed Russians top of the list of foreign buyers of London homes.

- Qatari investors own stakes in the Shard and Canary Wharf.
- In 1980, just 8% of the City of London was owned by non-national investors. In 2011, the proportion passed 50% (Figure 2).
- Many investment properties remain unoccupied, yet the owners change the identity of places.
- The trend in foreign ownership makes parts of London too expensive for local people.
- Many properties become gated compounds, turning public access into private spaces with security guards and CCTV.

Key
- Freehold properties
- Leasehold properties

▲ **Figure 2** *Properties owned by foreign investors in Central London in 2015*

Business ownership and identity

Some TNCs have altered the way of life, and even national identities, in some countries.

By the 1950s, the American United Fruit Company (now Chiquita) owned over 50% of all land in Honduras and 75% in Guatemala.

- It built and ran infrastructure.
- Plantations replaced small farms; waged work replaced subsistence farming; and homes, healthcare and education were all provided by the company.

In the UK, the Indian TNC Tata Steel owns the steel plant in Port Talbot, South Wales.

- It provides jobs and maintains the historical Welsh association with heavy industry.
- Tata claims that it benefits the Welsh economy and supports schools and local health, safety and environmental policies.
- In 2015, when cheap imported Chinese steel threatened Tata's profits, the company proposed closing the Port Talbot plant.
- Community futures are at risk when globalisation causes TNCs to change tack.

Ten-second summary

- Foreign-owned companies make 'Made in Britain' a complex idea.
- Westernisation is demonstrated through entertainment, retailing and the cultural influence of companies like Disney.
- Many properties and businesses are owned by non-nationals, which impacts on national identity.

Over to you

On sticky notes, write your top ten facts about why there are challenges to national identity. Stick them on your wall and test yourself on these every three days.

Student Book
See pages 260–263

You need to know:

- how strong nationalist movements seek independence
- about significant political tensions in the BRIC and other emerging nations
- that the role of the state is variable and national identity is not always strong.

Big idea

There are consequences of disunity within nations.

Nationalism versus globalisation

Both globalisation and nationalism are problematic.

- For some, globalisation means low wages, insecure employment, faceless organisations, stateless corporations and inequality.
- Rising nationalism – as well as calls for independence from regions within countries – has emerged alongside growing disillusionment with globalisation.
- However, some nationalist undercurrents have existed for many decades. For example, Catalonia has a long-held desire to become independent from the rest of Spain, while remaining in the EU.

Catalans demand change

Catalonia (Figure 1) is Spain's wealthiest region, producing 20% of Spain's wealth. Since 1978, Catalonia has enjoyed considerable levels of self-government.

- Many Catalans feel that they contribute more than their fair share to the nation's taxes and that they subsidise the poorer Spanish regions.
- By 2010, half of Catalonia's population wanted independence.

Calls for a vote on Catalan independence have grown.

- **2012** – 1.5 million Catalans demonstrated, demanding an independence referendum.
- **2014** – A mock referendum was held, with 80% of Catalan voters (on a 40% turnout) wanting separation from Spain.
- **2017** – The Catalan regional government held a full referendum about eventual independence, which Spain's constitutional court deemed to be illegal.

🔺 **Figure 1** *The location of Catalonia and its main city, Barcelona*

Catalonia and the EU

There are many European businesses in Catalonia. Catalan independence could mean renegotiation of trade deals, as well as changes to tax, regulations and currency. This would create challenges for businesses, similar to those arising from Brexit.

Rising tensions in emerging nations

Nationalism often has cultural and historical roots but it has been strengthened in places where globalisation has created tensions.

- The BRICs, and other emerging nations, have seen GDP grow significantly but they are grappling with the consequences of national divisions.
- For example, many Brazilians felt resentment at the cost of hosting the 2016 Olympic Games when Brazil was in the throes of a major recession, political strife and a public health crisis (the Zika virus).

Globalisation and inequality

Globalisation has produced winners and losers. Cheap loans have allowed many to buy cars or property, which, together with the falling prices of many consumer goods, has meant that many people 'feel' better off. But income inequality within most countries (measured by the Gini coefficient) has grown. Some of the emerging economies are the most unequal, whereas Sweden is the least unequal.

Inequality in South Africa

Globalisation has allowed South Africa's manufacturers to gain access to wider markets. However, the opening up of South Africa to global markets has also meant:

- taking on SAPs (see Section 5.9)
- accepting high levels of Foreign Direct Investment (FDI)
- the in-migration of overseas businesses.

Some South Africans have undoubtedly gained, but income inequalities between different ethnic groups in South Africa have actually increased (Figure 2), together with political tensions.

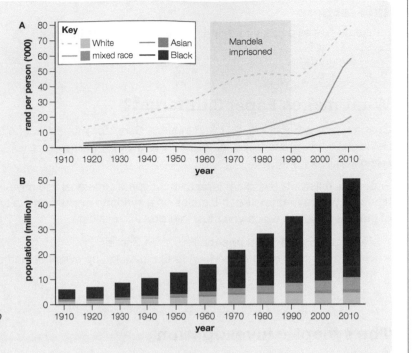

▶ **Figure 2** *Average income by ethnic group* **(A)** *and population by ethnic group* **(B)**

Failed states

A failed state exists where the political or economic system has become so weak that the national government is no longer in control.

- In failed states, some of the following apply: undemocratic government, social unrest, widespread poverty, human rights abuses, and poor education, health and welfare.
- A small elite may wield political and economic power.
- Any sense of national identity may be weak.

Syria

Civil war in Syria has contributed to its 'failure'.

- Civil war has raged since 2011, claiming nearly 500 000 lives and displacing millions as refugees.
- Aleppo, Syria's financial centre, has been reduced to ruins.
- Syria's infrastructure has disintegrated.
- The ongoing conflict involves up to 1000 armed opposition groups.
- The country is ethnically divided.

But the 'failure' of Syria is not new. Much of the problem has been caused by nepotism and corruption.

- Since 2000, the economy has largely been run by a group of entrepreneurs closely associated with the ruling family.
- During privatisation, public services and assets were transferred into the hands of political supporters of the government and those linked by family or clan.
- Meanwhile, the agricultural sector became neglected.

Ten-second summary

- Catalonia is a region that wants independence but also membership of the EU.
- Tensions in Brazil and South Africa have resulted from the costs and benefits of globalisation.
- Syria is an example of a failed state.

Over to you

Look at the 'Big idea'. Using examples, summarise what is meant by this statement.

In this section:

• you'll learn how to prepare for Paper 3 (the synoptic investigation).

What makes Paper 3 different?

Chapters in this book, together with those in *Geography for Edexcel A Level Year 1 and AS Revision Guide*, prepare you for Papers 1 and 2 by revising topics you've covered during your A level course.

Paper 3 is different. The exam is less about specific material you have learned and more about how well you have prepared. It focuses on a **synoptic evaluation** – a topic that synthesises parts of the specification, which you must interpret and evaluate.

• Paper 3 is based on an **unseen** Resource Booklet.
• In the exam, you need to read, understand, weigh up evidence and answer questions about the topic.

The synoptic investigation

The synoptic investigation is contained in a Resource Booklet about six to eight pages long.

• It will contain information that you probably won't have seen before about a geographical issue somewhere in the world. It's likely that you won't know much about this place. Don't worry – your knowledge of the place isn't being assessed. Instead, your ability to understand the issue *is* being assessed.
• The place is unlikely to be local in scale. Instead, it will be a broader area, such as a region (e.g. part of the USA or Canada), a country, or a group of countries (e.g. an economic union such as ASEAN). It could, unusually, even be a global issue. However, it is most likely to focus on particular places since this is where geographical issues are played out. The issue will be a real one, not fictional.
• To help you understand the issue, the Resource Booklet will contain text, maps, photos, diagrams and data. The exam assesses your ability to make sense of them.
• Towards the end of the Resource Booklet, you'll be asked to take an overview of the issue and link it to what you've learned in the course. This will be the focus of the longest exam question you'll answer in any of the papers – a 24-mark question.
• There won't be 'right' or 'wrong' answers. Instead, questions will assess your skills of analysis or evaluation. You'll be marked on how well you analyse data, argue, reason and make links to what you've studied.

The Resource Booklet will contain key words and concepts that you'll understand from pages 8 and 12. That's intentional – the examiners want you to make links between topics you've studied. This is called being **synoptic** – drawing together your knowledge, understanding and skills from your A level course.

What will the Resource Booklet be like?

Paper 3 will contain a Resource Booklet. Section 6.4 is an exam-style Resource Booklet. It is about Australia and focuses upon:

- Australia's economic background
- some of the issues that result from globalisation
- some of the environmental issues that Australia faces, e.g. its water insecurity
- its future direction.

Section 6.5 contains exam-style questions based on the Resource Booklet in Section 6.4. Section 6.7 contains a mark scheme for those questions.

Which topics are assessed?

The topic for the synoptic investigation will differ each year. But it will always be based on at least two of the **five compulsory topics**, and not about option topics. The five compulsory topics are:

- Tectonic Processes and Hazards (Chapter 1 in Book 1)
- Globalisation (Chapter 4 in Book 1)
- The Water Cycle and Water Insecurity (Chapter 1 in Book 2)
- The Carbon Cycle and Energy Security (Chapter 2 in Book 2)
- Superpowers (Chapter 3 in Book 2).

This doesn't mean you can't use knowledge learned in any option topics, such as:

- Glaciated Landscapes and Change or Coastal Landscapes and Change (Chapters 2 and 3 respectively in Book 1)
- Regenerating Places or Diverse Places (Chapters 5 and 6 respectively in Book 1)
- Health, Human Rights and Intervention or Migration, Identity and Sovereignty (Chapters 4 and 5 respectively in Book 2).

But it does mean that the examiners will not use these topics in the Resource Booklet and exam questions.

The exam

Paper 3 is a 2 hour 15 minute exam paper, worth 70 marks. It contributes 20% of the total towards your final A level grade. That means 70 marks in 135 minutes – less pressured than in Papers 1 and 2.

- However, remember that you'll need to read the content of the Resource Booklet carefully to make sense of it, as well as to plan your answers. A longer time allocation per question does not mean that you need to write a lot more – only that there is time to think, plan your answers and check your answers at the end.
- This paper demands considerable extended writing. There will be 4-mark questions. However there will also be two 8-mark questions (with 'Analyse' as the command word), together with an 18- and a 24-mark question (both with 'Evaluate' as the command word) – that's 58 marks from a total of 70 for the whole paper.
- Questions will demand use of your geographical skills, perhaps including manipulation of data (e.g. using correlation techniques such as Spearman's Rank). But questions of this nature will carry only about 4 marks.

In this section:
- you'll learn about how best to use the unseen Resource Booklet in the exam.

Exam-style Resource Booklet

Section 6.4 is an exam-style Paper 3 Resource Booklet, with exam-style questions in Section 6.5.

- It is based on Australia – its globalised economy, influence in ASEAN, water insecurity and environmental issues.
- You should see that most of the compulsory themes (see Section 6.1) are threaded through it.
- This should help you to understand how links can be made between the different topics you have studied.

More importantly, the Resource Booklet should help you to recognise that there is a wide spectrum of views about geographical issues and that these views are **contested**.

- You should also be aware that data in all their forms – e.g. statistics, photos, text – need treating with care and considerable suspicion.
- Look for the provenance of the sources you read – who collected them, their reliability and the likelihood that they will tend to put particular viewpoints across.

What examiners will look for

The exam will be divided into three sections (A–C) (Figure 1).

- Questions early in the paper will assess knowledge recall and skills in interpreting geographical information. These are intended to help you to gain an overview of the issue.
- Extended writing questions of up to 8 marks will ask you to analyse and explain trends that you see in the Resource Booklet.
- Finally, two extended essay-style questions, of 18 and 24 marks respectively, will probe your ability to evaluate material in the Resource Booklet. These questions (particularly the last one carrying 24 marks) will be based on a contested viewpoint, which you should be prepared to take on and discuss, using evidence to put forward your view.

To reach the top levels in the mark scheme, you must:

- show accurate geographical knowledge and understanding
- apply that knowledge and understanding to make logical and relevant connections/ relationships to material in the Resource Booklet
- interpret any data or material coherently, and support any arguments with evidence, so that you write rational, substantiated and balanced conclusions
- make valid judgements about the value and reliability of any data/evidence.

Section	Marks	Notes
A	12	• 1 question linking the Resource Booklet to previous knowledge and understanding, using 'Explain' as a command. It is normally worth 4 marks. • 1 question examining quantitative skills. It is normally worth 4 marks. • 1 question requiring some analysis. It is likely to use the command words 'Explain' or 'Suggest'. It is normally worth 4 marks.
B	16	• 2 x 8-mark mini-essays using the command word 'Analyse' and based directly on material in the Resource Booklet.
C	42	• 1 x 18-mark essay • 1 x 24-mark essay • Each question uses the command word 'Evaluate'. Each is based on analysis of the Resource Booklet (AO3) but also on your knowledge and understanding (AO1) and your ability to apply ideas (AO2).

⬆ *Figure 1* Question format in Paper 3

In this section:

• you'll learn about the synoptic themes that are the focus for Paper 3.

What are the synoptic themes?

Underlying the content of every topic in the Edexcel A Level Geography specification are three synoptic themes that 'bind' or glue the whole specification together. These themes are:

• **Players**
• **Attitudes and Actions**
• **Futures and Uncertainties**.

These are referred to throughout the student books and you should be clear about their meanings. Paper 3, the synoptic exam, will draw upon these themes and assess your ability to apply them.

Players

Players are those responsible for making decisions about people, the use of space and how these decisions are implemented. (Figure 1 on page 132 gives some examples of players in the compulsory topics.)

• They should not be confused with **stakeholders**, who are those who have a viewpoint about a contested issue, but are not decision-makers.
• Players are closely linked to political plans and strategies (e.g. the UK's economic transformation in the 1980s and its adoption of globalisation policies), specific plans (e.g. managing energy resources) or long-term programmes (e.g. responses to climate change).

Players can be categorised into three sectors: public, private and 'third'.

• **Private sector** includes businesses, ranging from small local companies to large TNCs.
• **Public sector** means organisations financed by public sources and functions, e.g. education, health, defence.
  – *Within* countries, public sector players range from small-scale (e.g. parish councils) to regional (e.g. county councils) to national government.
  – *Beyond* individual countries, they include IGOs and economic unions such as the EU.
The interaction between private and public sector players is critical in decision-making.
• **Third sector** players include pressure groups (e.g. Greenpeace), NGOs (e.g. Oxfam) and political **think-tanks** (highly significant in terms of political influences and ideas).

Attitudes and Actions

Attitudes are the viewpoints that decision-makers and stakeholders have about economic, social, environmental or political issues. **Actions** are the ways in which they try to achieve what they want. (Figure 1 on page 132 gives some examples of attitudes and actions in the compulsory topics.)

• Attitudes and actions are therefore linked to players.
• They are important, because players with certain attitudes about one issue (e.g. pro-globalisation) *may* have similar views about other issues (e.g. anti-climate change).
• The media play a role in establishing attitudes about particular issues.

Futures and Uncertainties

This is about 'big questions' such as:

• Can the world provide people with safe water to drink?
• How far will climate change play a part in any future decision-making?

(Figure 1 gives some examples of futures and uncertainties in the compulsory topics.)

The futures that different players envisage is strongly linked to their attitudes. These attitudes may be:

• **Business as usual**, i.e. let things stay as they are, or 'do nothing'. For example: 'Private companies should be allowed to decide energy futures, letting market forces drive energy policy.'
• More **sustainable strategies**, such as radical action in managing climate change. For example: 'Governments should encourage pro-renewable energy policies.'

	Players	Attitudes and Actions	Futures and Uncertainties
Tectonic Processes and Hazards (Chapter 1 in Book 1)	• Local and national governments • The roles of scientists, planners, NGOs and engineers in hazard management and prediction		
Globalisation (Chapter 4 in Book 1)	• WTO, IMF, World Bank, EU, ASEAN, governments and their policies towards economic liberalisation, attracting FDI and TNCs • Opportunities for disadvantaged groups	• Actions taken in support of and against globalisation • Environmental movements arising from the impacts of globalisation • Viewpoints in favour of and against immigration • Actions of NGOs and pressure groups	• Environmental consequences of resource consumption
The Water Cycle and Water Insecurity (Chapter 1 in Book 2)	• The role of planners in, for example, managing land use or water supply • Various players in trans-boundary and internal conflicts	• Contrasting attitudes to water supply (e.g. comparing smart irrigation with water recycling schemes and mega-dams)	• Projections of future drought/flood risk • Projections of future water scarcity
The Carbon Cycle and Energy Security (Chapter 2 in Book 2)	The roles of: • TNCs, OPEC, consumers, government • businesses in involved in exploiting resources versus environmental groups and affected communities	• Attitudes of global consumers to environmental issues • Attitudes of different countries, TNCs and people	• Uncertainty of global projections
Superpowers (Chapter 3 in Book 2)	The roles of: • TNCs in maintaining power and wealth • powerful countries as a 'global police' • emerging powers in geopolitical change	• Actions and attitudes of global IGOs and different countries towards globalisation • Attitudes of different players in relation to resources • Contrasting cultural ideologies between different countries and players	• Uncertainty over future power structures as the balance between superpowers changes

 Figure 1 *The synoptic themes in the five compulsory topics. This should help you to think about the links between the three themes*

In this section:

- You'll learn about the characteristics of Australia and the issues it faces in its economic development.
- Section 6.4 is an exam-style Resource Booklet as you would find in Paper 3, the synoptic exam.
- Section 6.5 contains exam-style questions about this Resource Booklet, of the kind you could meet in a Paper 3 exam.
- Section 6.6 contains information on how Paper 3 is assessed.

Big idea

You are advised to read all sections before attempting any questions, both here and in the actual exam.

Section A: Australia – an overview

- Australia is the world's sixth largest country (Figure 1). With 7.6 million km², it is 30 times bigger than the UK and similar in size to the USA, excluding Alaska.
- Australia is the world's driest inhabited continent and is vulnerable to drought, El Niño and the challenges of climate change. Water is precious and rainfall is low over much of the country (Figure 2).
- In recent decades, Australia's economy has grown rapidly and it is now one of the world's highest-income, globalised economies.
- Its population is growing, largely due to a high rate of inward migration. Rapid urbanisation is occurring, especially in the largest cities, where young people are attracted by city lifestyles.
- In 1901, 61% of Australians lived in rural areas, small towns and smaller regional cities. By 2007, this had fallen to less than 17% and is forecast to fall to 7% by 2050.
- Australia's environmental challenges include water shortages, soil degradation, floods, droughts and bushfires.
- Its economic challenges include an ageing population, pressure on infrastructure (due to rapid population growth) and an economy that is highly dependent upon markets in China and South-East Asia.

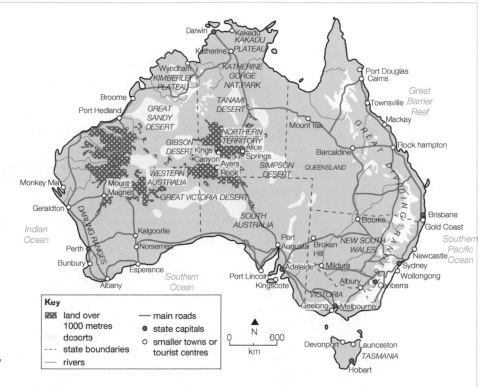

Figure 1 Australia's key features

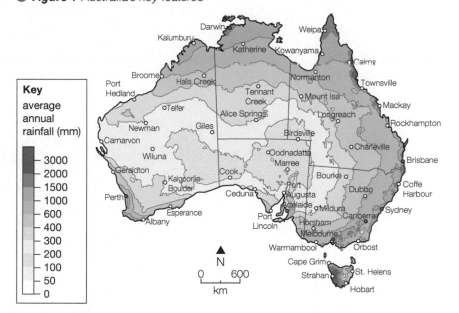

Figure 2 Rainfall distribution in Australia

Section B: Australia's population

- In 2017, Australia's population was 24.3 million (Figure 3), an increase of 14% since 2009.
- Its annual population increase was 1.03%, one of the highest of any developed country (Figure 4).
- Australia's development has historically been concentrated in a 2500 km belt along its south-east coast, between Brisbane and Melbourne in the states of Queensland, New South Wales and Victoria (Figure 5).
- Development of the interior has mostly been around water sources (e.g. rivers or aquifers) and mining resources.

Date	Population (millions)
1901	3.8
1921	5.5
1941	7.1
1961	10.5
1981	15.0
2001	19.5
2017	24.3

⬆ **Figure 3** Australia's population 1901–2017

- Birth rate: 12.1 per 1000
- Death rate: 7.3 per 1000
- Rate of natural increase: 4.8 per 1000
- Fertility rate (2016): 1.8 children per woman
- Net migration rate: 5.5 per 1000
- HDI score: 0.939
- Life expectancy: 84.5 (for Aboriginal Australians, life expectancy is 10.6 years less for men and 9.5 for women)
- Infant mortality rate: 4.3 per 1000 live births
- Urban population: 89.7% of total population

⬆ **Figure 4** Australia's key population indicators (2017 unless indicated otherwise)

	Population (number)	Population (% of country total)	Area (km²)	Area (% of country total)
New South Wales	7 861 100	31.96%	800 641	10.41%
Victoria	6 323 600	25.71%	227 416	2.96%
Queensland	4 928 500	20.04%	1 730 647	22.50%
Australian Capital Territory*	410 300	1.67%	2 358	0.03%
South Australia	1 723 500	7.01%	983 482	12.79%
Northern Territory	246 100	1.0%	1 349 129	17.54%
Western Australia	2 580 400	10.49%	2 529 875	32.9%
Tasmania	520 900	2.12%	68 401	0.89%

⬆ **Figure 5** Distribution of population by state, 2017. (* Australian Capital Territory is the small area around Canberra, Australia's capital city)

	New South Wales	Victoria	Queensland	Western Australia	South Australia	Tasmania	Australian Capital Territory	Northern Territory	Total
Row 1 Observed population (O) share %	31.96	25.71	20.04	10.49	7.01	2.12	1.67	1.00	100
Row 2 Expected (E) share %	12.50	12.50	12.50	12.50	A	12.50	12.50	12.50	100
O–E	19.46	13.21	B	–2.01	–5.49	–10.38	–10.83	–11.50	
O–E²	378.69	174.50	56.85	4.04	C	107.74	117.29	132.25	
(O–E)²/E	30.30	13.96	4.55	0.32	D	8.62	9.38	10.58	**80.12**

⬆ **Figure 6** The actual (or Observed – O) population share of Australia's eight states and territories (Row 1) compared to how this population would be distributed if all eight states and territories had equal (or Expected – E) share of the population (Row 2)

Section C: Australia's economy and trade

- GDP (2016): AU$1.2 trillion
- GDP per capita: US$49 900, an increase of over 50% since 2008, and over 230% since 1990
- Employment by occupation: agriculture 3.6%; mining and industry 26.1%; services 70.3%
- Exports: iron ore, coal, gold, natural gas, aluminium ores and concentrates, wheat, meat, wool, alcohol (wine)
- Export markets (2016): China 30.5%, Japan 12.4%, USA 6.5%, South Korea 6.1%
- Half of the value of Australia's exports comes from mineral products (Figure 9) from places such as the Pilbara region in the north of Western Australia (Figures 8 and 9).
- Imports (2016): motor vehicles, refined petroleum, telecommunication equipment and parts; crude petroleum, medicines, goods vehicles, gold, computers.

▲ **Figure 8** *The Pilbara region in Western Australia – much of Australia's iron ore exports are carried in trains such as this*

▲ **Figure 7** *Key economic indicators for Australia. All data are for 2017 unless indicated otherwise*

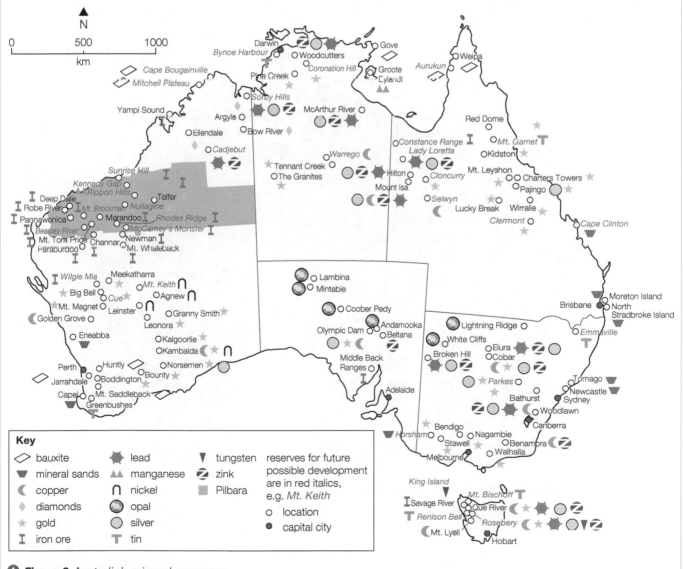

▲ **Figure 9** *Australia's mineral resources*

Section D: Australia's water resources and security

There has been a debate for a long time about how many people Australia can support.

- Australia has an increasingly fragile environment. A strong environmental lobby believes that Australia can only produce enough food and resources for a limited number of people.
- Water is increasingly scarce given the demands made on it (Figure 11).
- Water schemes such as the Snowy Mountains scheme in the 1950s and 1960s (Figure 10) were designed to provide water for the cities of south-eastern Australia. Those schemes are no longer sufficient to meet demand.
- Much agriculture in the modified grazing pastures, irrigated cropping and dryland cropping regions shown in Figure 12 (on page 137) is fed by spray irrigation (Figure 13 on page 137) drawn from aquifers and artesian water from underground. Aquifer levels are falling.
- There is a belief among some Australians that the population should be reduced to prevent further degradation of Australia's unique environments. They believe that food and wine production in Australia have each caused major environmental problems.

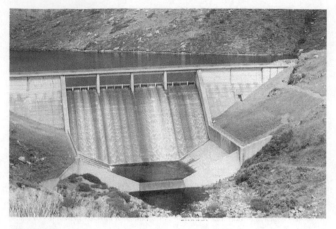

⬣ **Figure 10** *The Snowy River scheme, typical of many water storage schemes, in the mountains of Victoria and New South Wales. These provide water for cities such as Sydney and Melbourne, and for agriculture*

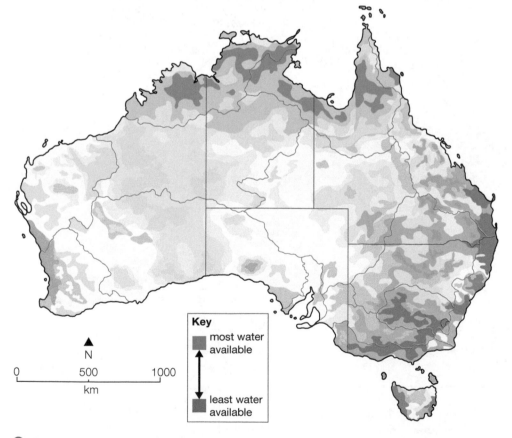

Key

■ most water available

↕

■ least water available

N

0   500   1000
km

⬣ **Figure 11** *Water scarcity in Australia*

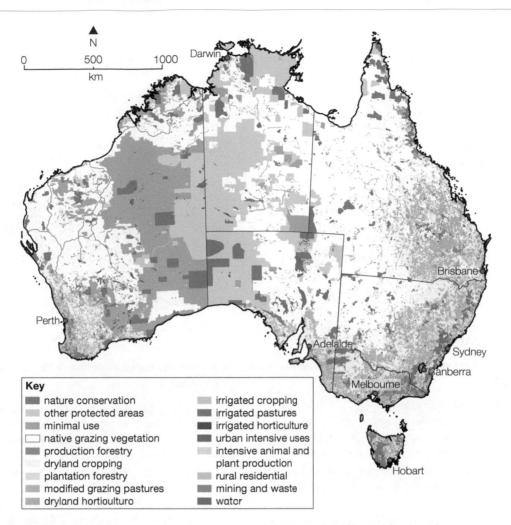

Key

- nature conservation
- other protected areas
- minimal use
- native grazing vegetation
- production forestry
- dryland cropping
- plantation forestry
- modified grazing pastures
- dryland horticulture
- irrigated cropping
- irrigated pastures
- irrigated horticulture
- urban intensive uses
- intensive animal and plant production
- rural residential
- mining and waste
- water

⬆ **Figure 12** *Land use in Australia, 2013. Some of the greatest environmental impacts of agriculture can be seen in the modified grazing pastures, irrigated cropping and dryland cropping areas shown on this map. In these areas, agriculture is only possible on a large scale using irrigation*

⬆ **Figure 13** *Spray irrigation in fields near the town of Parkes in rural New South Wales*

Section E: Australia's environmental fragility

There has been a debate for a long time about how many people Australia can support.

- Australia has an increasingly fragile environment. A strong environmental lobby believes that Australia can only produce enough food and resources for a limited number of people.
- There is a belief that the population should be reduced to prevent further degradation of Australia's unique environments.
- They believe that food and wine production in Australia has caused major environmental problems.
- Soil erosion (Figure 14) has resulted from clearance for agriculture in some drier areas of the country. Some people fear that further expansion of agriculture could lead to degraded soils in many parts of Australia.

△ **Figure 14** *Soil erosion in rural South Australia, the result of land clearance in the late 19th century*

What's the cause?

Extract 1

It's wrong to think that Australia could expand its population to 50 million by 2050. This would need the construction of a city the size of Sydney every seven years and an annual immigration rate of 450 000. It is as if people who believe this have learned nothing about the pressures on Australia's environment.

(Clive Hamilton, Professor of Public Ethics at Charles Sturt University, Canberra)

Extract 2

Land degradation has sometimes been attributed to population levels. But … Australian soils are in fact feeding more than 56 million people (both within and outside Australia), and provide wool and cotton for even more people. Moreover, damage done to the soils was done by very small populations: the colonial settlers who cleared the land and the farmers (now less than 5 per cent of the total population) who still clear the land and sometimes cause further soil erosion. Similarly, mining also provides the needs of a wider population. The National Population Council argues that the 'environmental impact of any industry which exports a very large proportion of its output is therefore weakly related to domestic population needs and requirements'.

(Professor Sharon Beder, The Nature of Sustainable Development (2nd edition), Scribe 1996)

Extract 3

… Almost all major environmental battles will be lost if we fail to win the population battle now. The results of high population growth have been dead river systems, near-permanent water shortages, increasing pollution, a surge in imports and skyrocketing foreign debt, reduced per capita value of our mineral wealth and exports, expensive rebuilding of our cities and infrastructure, loss of limited arable farmland to housing, crowding of our coastal towns and resorts, loss of native species and wildlife, urban congestion, and loss of open spaces for our children to play, just to name a few. What is all this in aid of? The great Australian dream has turned into a nightmare of haves and have-nots. If Australians knew that it would have led to all this, they would have turfed governments pushing high population growth out of office.

(William Bourke, The Sydney Morning Herald, 18 February 2010)

△ **Figure 15** *Three viewpoints about Australia's development*

In this section:
- you'll find advice on, and examples of, exam-style questions.

Preparing for the exam-style questions

Because this is a practice, Activities **A** and **B** below will help you to prepare in advance for the exam-style questions that follow. Each activity can be done in small groups.

Activity A

Make links between Australia, its people, its economic indicators, its water security and its environmental problems.

Consider how the synoptic teaching topics on which this Resource Booklet is based (Globalisation, Water cycle and water insecurity, Carbon cycle and energy security, and Superpowers) link to Australia and its issues.

Activity B

Consider the roles of Players, their Attitudes and Actions, and the Futures and Uncertainties that Australia faces. Who might be the decision-makers in issues about Australia's future population, its resource management, water security and environmental management?

Exam-style questions

Instructions:

- Answer ALL questions.
- Use the Resource Booklet (Section 6.4) and your own knowledge and understanding from across your two-year course of study to answer these questions.

1 Study Section A of the Resource Booklet. Explain the relationship between economic development and water supply. *(4 marks)*

2 a Figure 6 in the Resource Booklet shows data on population distribution for Australia that has been prepared to carry out a chi-square calculation. In the table, the actual (or Observed – O) population share of Australia's eight states and territories is shown in Row 1. Row 2 shows the percentage share of how this population would be distributed if all eight states and territories had an equal (or Expected – E) population share.
Using the data in Figure 6, find the missing data for boxes A, B, C and D. *(4 marks)*

b Explain how you would use the result (80.12) to analyse whether the population distribution of Australia had occurred by chance. *(4 marks)*

3 a Study Section B of the Resource Booklet. Analyse the extent to which Australia's population is typical of that of a developed economy. *(8 marks)*

b Study Figure 7 in Section C of the Resource Booklet. Analyse the extent to which Australia's economic indicators are typical of those of a developed economy. *(8 marks)*

4 Study Section D of the Resource Booklet. Evaluate the role of water in Australia's economic progress. *(18 marks)*

5 Evaluate the constraints that developed countries such as Australia may face in achieving future economic growth. *(24 marks)*

Total 70 marks

In this section:
* you'll learn the broad principles of how Paper 3 is assessed and marked.

Understanding assessment objectives

Examiners decide how they will assess you, using **Assessment Objectives** (AOs). There are three at A level:

* **AO1 Knowledge and understanding** – worth 34% of the total mark for A level. These questions test what you know and have learned throughout the course.
* **AO2 Application of understanding** – e.g. in interpreting, analysing and evaluating geographical information and issues, worth 40% of the total mark for A level. These questions test how you interpret situations or data that you are presented with. They are the questions that assess your ability to put together a coherent argument, to make judgments and to apply what you know to develop an evidenced case.

> ▶ **Figure 1** *Assessment Objectives (AOs) for Edexcel A Level Geography*

* **AO3 Geographical skills** – e.g. in handling and processing information in investigating geographical questions and issues (e.g. interpretation, analysis and evaluation of data and evidence), constructing arguments and drawing conclusions. This AO is worth 26% of the total mark for A level. These questions test your skills.

The distribution of AOs in Paper 3 is very different from Papers 1 and 2 (Figure 1).

Paper	Assessment Objectives (AOs)			Total for all AOs
	AO1	AO2	AO3	
Paper 1	13%	15.75%	1.25%	30%
Paper 2	13%	15.75%	1.25%	30%
Paper 3	5.5%	6%	8.5%	20%
NEA	2.5%	2.5%	15%	20%
Total	**34%**	**40%**	**26%**	**100%**

Understanding how Paper 3 is assessed

Because Paper 3 is based on an unseen Resource Booklet, the most obvious assessment objective to be assessed is your ability to interpret, grasp and understand what's in the Resource Booklet – your geographical skills (**AO3**).

* You'll notice that AO3 is the largest assessment objective assessed in Paper 3 (8.5% of the total of 20% for Paper 3 as a whole – Figure 1). In contrast, AO3 is very minor in Papers 1 and 2.

However, your performance in Paper 3 will also depend on the two other assessment objectives.

* You'll need to apply what you see in the Resource Booklet to what you have learned across the course. This means that you are being assessed in **AO2** – worth 6% of the 20% weighting for Paper 3 as a whole. So when you are making judgments about water scarcity in Australia (Section D in the Resource Booklet, Section 6.4), you'll be applying what you learned in the topic on The Water Cycle and Water Insecurity.
* You'll also need to demonstrate particular knowledge and understanding from topics that you have learned. This means that you are being assessed in **AO1** – worth 5.5% of the 20% weighting for Paper 3 as a whole. So you should use examples that you have learned in relevant topics in your answers, particularly in the last two questions. You might use examples from the topic on Globalisation, for example, in discussing Australia's economy in question 3b (Section 6.5) and how typical it is of developed economies.

How Paper 3 is marked

Like the rest of A Level Geography:

- Questions carrying 4 marks or fewer are point-marked – you are awarded 1 mark for every correct point that you make or the way in which you develop it.
- Questions carrying 6 marks or more are level-marked, based on criteria that examiners use to judge the overall quality of an answer.

In Paper 3, you can expect most of the questions, therefore, to be level-marked.

- The question format is likely to be very similar from one year to another, with questions using particular command words (Figure 2).
- The question style is also likely to be similar each year (Figure 3).

Mark tariff	4	8	18	24
Calculate	*			
Draw/Plot	*			
Explain	*	*		
Analyse (used only in Paper 3)		*		
Evaluate			*	*

▲ **Figure 2** *Command words used in question setting and the marks allocated to each one in Paper 3*

Notice in Figure 3 how the assessment objectives are assessed.

- Some questions, like questions 1–3, assess single assessment objectives. For example, in the exam-style questions in Section 6.5, question 1 is assessing AO1 – you would be expected to know the relationship between economic development and water supply.
- Extended paragraphs, meanwhile, assess more than one assessment objective. For example, in question 3b (Section 6.5), your ability to analyse the extent to which Australia's economic indicators are typical of those of a developed economy depends upon:
  - your knowledge of what the indicators of developed countries are like (AO1)
  - your ability to interpret the data in Figure 7 in the Resource Booklet (AO3).

Therefore, the 8 marks for question 3b are split – 4 marks for AO1 and 4 for AO3.

- Longer essays develop this trend even more. 18- and 24-mark questions in Paper 3 assess all three assessment objectives.
  - 18-mark questions are split 3 marks for AO1 – 9 for AO2 and 6 for AO3.
  - 24-mark questions are split 4 marks for AO1 – 12 marks for AO2 and 8 for AO3.

In each case, you need to be able to develop an argument (AO2), based on your own knowledge and understanding (AO1) and the use of data from the Resource Booklet (AO3).

Section	Question	Marks	Question type	Notes
A	Q1–Q3	12	Short open responses	• 1 AO1-based 'Explain...' question – usually 4 marks • 1 question examining skills (AO3) – usually 4 marks • 1 question requiring analysis (AO3) – usually 4 marks
B	Q4 and Q5	16	Extended paragraphs	• 2 x 8-mark paragraphs, using command word 'Analyse', based on data in the Resource Booklet (a mix of AO1 and AO3)
C	Q6 and Q7	42	Longer essays	• 1 x 18-mark essay based on AO3 'reading' of the Resource Booklet but also assessing AO1 and AO2 • 1 x 24-mark essay based on AO3 'reading' of the Resource Booklet but also assessing AO1 and AO2

▲ **Figure 3** *Question styles and AO usage in question setting in Paper 3*

1 **Study Section A of the Resource Booklet. Explain the relationship between economic development and water supply.**

(4 marks)

This question targets AO1 (knowledge and understanding).

Instructions for marking

1 mark is awarded for identifying a correct factor in the relationship between economic development and water supply (up to a maximum of 2 marks), with a further 2 marks for expansion. For example:

- Water is essential to farming in order to feed an urban population (1):
    - because much economic growth is urban and urban populations need to be fed (1), OR
    - for domestic water supply in urban areas (1), OR
    - because, as the rural population declines, people migrate to cities for jobs (1).
- Irrigation systems help to counter-balance water deficits (1) that might otherwise prevent food production or water supply (1).
- Water is essential to many energy-producing / industrial processes (1) so they can generate energy for industrial or domestic consumption (1).

2a **Figure 6 in the Resource Booklet shows data on population distribution for Australia that has been prepared to carry out a chi-square calculation. In the table, the actual (or Observed – O) population share of Australia's eight states and territories is shown in Row 1. Row 2 shows the percentage share of how this population would be distributed if all eight states and territories had an equal (or Expected – E) population share.**

Using the data in Figure 6, find the missing data for boxes A, B, C and D.

(4 marks)

This question targets AO3 (using geographical skills) for 4 marks.

Instructions for marking

1 mark is awarded for each of the following correct answers:

- A = 12.5
- B = 7.54
- C = 30.14 (allow 30.1)
- D = 2.41 (allow 2.4).

2b **Explain how you would use the result (80.12) to analyse whether the population distribution of Australia had occurred by chance.**

(4 marks)

This question targets AO3 (using geographical skills) for 4 marks.

Instructions for marking

1 mark is awarded for identifying any correct part of the process in exploring the relationship, up to a maximum of 4 marks. For example:

- Using a significance table (1) to assess whether the relationship between actual and expected population distribution had occurred by chance (1).
- Plotting the result (80.12) on the graph against the degrees of freedom (1) and reading off the significance line (e.g. 99%) (1), which would show how far the result was caused by chance.
- The strongest result would occur if it was less than 0.1% (1), i.e. indicating that there was less than 0.1% probability that Australia's population distribution occurred by chance (1).

3a Study Section B of the Resource Booklet. Analyse the extent to which Australia's population is typical of that of a developed economy.

(8 marks)

> This question targets AO1 (knowledge and understanding) for 4 marks and AO3 (using geographical skills) for 4 marks.

Instructions for marking

The general marking guidance for AO1 and AO3 below lists points that gain credit; the list is not prescriptive and you do not need to include all the points. The criteria in the level-based mark scheme for Question 3b (on page 144) are also applied.

General marking guidance for AO1

- Only a few developed countries (or HICs) have such rapid population growth (the UK is another example).
- Most developed countries have a slower population growth than Australia. Australia is typical of 'successful' countries that have pursued a high immigration rate.
- Most developed countries have a higher death rate than Australia, partly the result of its young population.
- Australia's fertility rate is typical of most developed countries – below replacement level.
- Its HDI, life expectancy and infant mortality are typical of most developed countries or are even higher than most.
- Many developed countries have a higher population density.
- Although most developed countries have a high percentage urban population, Australia is among the world's highest.

General marking guidance for AO3

Identifies and contextualises:

- the trend in population total and its recent acceleration
- the data in Figure 4 in terms of low natural increase but high net migration rate, as well as data on birth rates (low), death rate (very low) and HDI score (very high)
- the inequalities for Aboriginal Australians
- Australia's very low population density.

3b Study Figure 7 in Section C of the Resource Booklet. Analyse the extent to which Australia's economic indicators are typical of those of a developed economy.

(8 marks)

> This question targets AO1 (knowledge and understanding) for 4 marks and AO3 (using geographical skills) for 4 marks.

Instructions for marking

The general marking guidance for AO1 and AO3 below lists points that gain credit; the list is not prescriptive and you do not need to include all the points. The criteria in the level-based mark scheme (on page 144) are also applied.

General marking guidance for AO1

- Australia's GDP is high and GDP per capita very high (among the world's highest).
- Most developed countries have a slower economic growth than Australia's.
- Its employment structure is typical of highly globalised countries that have pursued the development of a knowledge economy, particularly in the percentage contribution by the services sector.
- Few developed countries rely on primary products to the extent that Australia does.
- Its imports are mainly manufactured goods.
- Its markets are predominantly in Asia with only a low percentage of exports to the USA and none to the EU.

General marking guidance for AO3

Identifies and contextualises:

- the typicality of GDP and GDP per capita among highly developed economies
- the regional nature of Australia's exports and its reliance on primary products (perhaps with a comment about the long-term impact of that if demand from China is reduced or about the finite nature of its mineral products)
- its imports are rather like those of a developing or emerging economy.

Level	Mark	Descriptors for Questions 3a and 3b
0	0	No rewardable material
1	1–3	• Demonstrates isolated elements of geographical knowledge and understanding, some of which may be inaccurate or irrelevant. (AO1) • Investigates the question/issue to produce a limited analysis of data/evidence, making few connections to geographical ideas. (AO3)
2	4–6	• Demonstrates geographical knowledge and understanding that is mostly relevant but may include some inaccuracies. (AO1) • Critically investigates the question/issue to produce an analysis of data/evidence, making some logical connections to geographical ideas, which are mostly relevant. (AO3)
3	7–8	• Demonstrates accurate and relevant geographical knowledge and understanding throughout. (AO1) • Critically investigates the question/issue to produce a coherent analysis of data/evidence, making logical connections to relevant geographical ideas. (AO3)

---

4 **Study Section D of the Resource Booklet. Evaluate the role of water in Australia's economic progress.**

(18 marks)

> This question targets AO1 (knowledge and understanding) for 3 marks, AO2 (application) for 9 marks and AO3 (using geographical skills) for 6 marks.

Instructions for marking

The general marking guidance for AO1, AO2 and AO3 below lists points that gain credit; the list is not prescriptive and you do not need to include all the points. The criteria in the level-based mark scheme on page 145 are also applied.

General marking guidance for AO1

Points should focus on the importance of water in economic development and some of the issues that arise, e.g.

- surpluses and deficits within the hydrological system and how a dry country such as Australia faces water deficit issues
- the links between water consumption and economic development
- the potential for water insecurity among some western developed economies
- the different strategies for adopting more sustainable approaches to water management.

General marking guidance for AO2

Arguments should include a range of potential impacts of the role of water in Australia's economic progress, e.g.

- the essential nature of water and its role in economic development
- how little of Australia has development potential without water; even mining puts pressure on dry environments
- the debate about whether water will be the issue that limits the Australian economy in future
- the unsustainable nature of water use in Australia and its impact on where urban and economic growth might occur in future.

General marking guidance for AO3

Data from a range of the images in the Resource Booklet should be used to support arguments, including:

- the fragility of Australia's natural and its economic environment caused by water deficits, and the pressure on different land uses (Figure 12)
- the concentration of population in three states (Figures 1 and 5) and the demands on natural ecosystems and water to supply those populations with water and (via irrigation) food (e.g. the dam in Figure 10)
- the geographical pattern of water scarcity in Australia (Figure 11), coinciding with areas of highest population density
- impacts of irrigation and the wasteful use of water by spray irrigation (Figure 13).

Level	Mark	Descriptor
0	0	No rewardable material
1	1–6	• Demonstrates isolated elements of geographical knowledge and understanding, some of which may be inaccurate or irrelevant. (AO1) • Applies knowledge and understanding of geographical information/ideas, making limited and rarely logical connections/relationships. (AO2) • Applies knowledge and understanding of geographical information/ideas to produce an interpretation with limited relevance and/or support. (AO2) • Applies knowledge and understanding of geographical information/ideas to produce an unsupported or generic conclusion, drawn from an argument that is unbalanced or lacks coherence. (AO2) • Makes superficial judgements about the value and reliability of quantitative and qualitative data/evidence. (AO3) • Investigates the question/issue to produce a limited interpretation of quantitative and qualitative data/evidence, but lacks meaningful connections to geographical ideas from across the course of study. (AO3)
2	7–12	• Demonstrates geographical knowledge and understanding that is mostly relevant but may include some inaccuracies. (AO1) • Applies knowledge and understanding of geographical information/ideas to find some logical connections/relationships. (AO2) • Applies knowledge and understanding of geographical information/ideas to produce a partial but coherent interpretation that is supported by some evidence. (AO2) • Applies knowledge and understanding of geographical information/ideas to reach a conclusion, partially supported by an argument that may be unbalanced or partially coherent. (AO2) • Makes some valid judgements about the value and reliability of quantitative and qualitative data/evidence. (AO3) • Investigates the question/issue to produce an interpretation of quantitative and qualitative data/evidence, making some meaningful connections to geographical ideas from across the course of study. (AO3)
3	13–18	• Demonstrates accurate and relevant geographical knowledge and understanding throughout. (AO1) • Applies knowledge and understanding of geographical information/ideas to find fully logical and relevant connections/relationships. (AO2) • Applies knowledge and understanding of geographical information/ideas to produce a full and coherent interpretation that is supported by evidence. (AO2) • Applies knowledge and understanding of geographical information/ideas to come to a rational, substantiated conclusion, fully supported by a balanced argument that is drawn together coherently. (AO2) • Makes valid judgements about the value and reliability of quantitative and qualitative data/evidence throughout. (AO3) • Critically investigates the question/issue to produce a coherent interpretation of quantitative and qualitative data/evidence, making meaningful connections to relevant geographical ideas from across the course of study throughout the response. (AO3)

5 Evaluate the constraints that developed countries such as Australia may face in achieving future economic growth.

(24 marks)

This question targets AO1 (knowledge and understanding) for 4 marks, AO2 (application) for 12 marks and AO3 (using geographical skills) for 8 marks.

Instructions for marking

The general marking guidance for AO1, AO2 and AO3 below and on page 146 lists points that gain credit; the list is not prescriptive and you do not need to include all the points. The criteria in the level-based mark scheme on page 146 are also applied.

General marking guidance for AO1

Points should focus on some of the constraints facing Australia's future economic development, e.g.

- sustainable development (e.g. Brundtland), especially in the context of the resources in Section E (Australia's environmental fragility)
- understanding how far globalisation has helped Australia's economic development and how far it has become a services-led economy
- how far primary products can offer sustainable growth – or whether the bonanza of selling raw materials to China to service economic growth in the past 30 years offers a sustainable way forward
- the challenges brought by physical distance from the rest of the world and the cost implications caused by travel and distance
- understanding the impacts of mineral exploitation, including whether minerals offer a model for sustainable growth.

General marking guidance for AO2

Arguments should include a range of potential constraints upon Australia's future economic progress, e.g.

- theory about, for example, the global shift and how far Australia is ideally located geographically to generate future economic growth
- the economic benefits of the service economy, globalisation, FDI and primary exports, and how far Australia has relied upon Asian growth rather than develop a sustained balanced economy
- environmental issues that may lie ahead, as shown in Section E (soil fragility, water scarcity)
- the debate about just how large Australia's population should be or could be – the 'limits to growth' versus arguments put forward in Extract 2 on page 138.

General marking guidance for AO3

Data from a range of images in the Resource Booket should be used to support arguments, including:

- the importance of water scarcity as a future challenge
- the compromise between economic growth (including mineral exploitation) and some of the impacts upon the environment
- data to show the stresses derived from water scarcity
- the impacts of irrigation and the ways in which issues such as water extraction compromise sustainable development.

Level	Mark	Descriptor
0	0	No rewardable material
1	1–6	• Use the criteria for Level 1 for Question 4 above.
2	7–12	• Demonstrates geographical knowledge and understanding that is occasionally relevant and may include some inaccuracies. (AO1) • Applies knowledge and understanding of geographical information/ideas with limited but logical connections/relationships. (AO2) • Applies knowledge and understanding of geographical information/ideas to produce a partial interpretation that is supported by some evidence but has limited coherence. (AO2) • Applies knowledge and understanding of geographical information/ideas to come to a conclusion, partially supported by an unbalanced argument with limited coherence. (AO2) • Makes some valid judgements about the value and reliability of quantitative and qualitative data/evidence. (AO3) • Investigates the question/issue to produce an interpretation of quantitative and qualitative data/evidence, making few connections to geographical ideas from across the course of study, which may not be meaningful. (AO3)
3	13–18	• Demonstrates geographical knowledge and understanding that is mostly relevant but may include some inaccuracies. (AO1) • Applies knowledge and understanding of geographical information/ideas to find some logical connections/relationships. (AO2) • Applies knowledge and understanding of geographical information/ideas to produce a partial but coherent interpretation that is supported by some evidence. (AO2) • Applies knowledge and understanding of geographical information/ideas to come to a conclusion, partially supported by an argument that may be unbalanced or partially coherent. (AO2) • Makes some valid judgements about the value and reliability of quantitative and qualitative data/evidence. (AO3) • Investigates the question/issue to produce an interpretation of quantitative and qualitative data/evidence, making some meaningful connections to geographical ideas from across the course of study. (AO3)
4	19–24	• Use the criteria for Level 3 for Question 4 above.

Glossary

This glossary includes key words and terms related to your Edexcel A Level Geography specificaton.

A

adaptation strategies – strategies designed to prepare for and reduce the impact of events

afforestation – the re-planting of trees when deforestation has occurred, or establishing forests on land not previously forested

aid dependency – the level to which a country cannot perform many of the basic functions of government without overseas aid

albedo – the amount of heat that is reflected by the Earth

antecedent moisture – water from one storm that has not drained away before more rain arrives

apartheid – the enforced segregation of people by skin colour or ethnicity. This policy was used in South Africa between 1948 and 1991

aquaculture – the breeding and harvesting of aquatic animals and plants

aquifer – an underground reservoir most commonly formed in rocks such as chalk and sandstone

Arctic amplification – the phenomenon where the Arctic region is warming twice as fast as the global average

arithmetic scale – increasing by a standard value each time – e.g. 2, 4, 6, 8 etc

assimilation – the gradual integration of an immigrant group into the lifestyle and *culture of the host country

asylum seekers – people who are fleeing to another country and applying for the right to international protection

austerity – the policy of reducing government spending and debt

B

base flow – also known as *groundwater flow – slow-moving water that seeps into a river channel through rocks

bio-geochemical carbon cycle – the continuous transfer of carbon from one *store to another, through the processes of photosynthesis, respiration, decomposition and combustion

biological carbon pump – where phytoplankton in the oceans sequester carbon dioxide through the process of photosynthesis – pumping it out of the atmosphere and into the ocean *store

biological decomposers – organisms such as insects, worms and bacteria which feed on dead plants, animals and waste

biologically derived carbon – carbon which is stored in shale, coal and other sedimentary rocks

BRICs – the collective term for Brazil, Russia, India and China (and, latterly, South Africa) which were predicted (by writer Richard Scase in 2000) to show rapid economic growth

C

carbon capture and storage (CCS) – the technological 'capturing' of carbon emitted from power stations

carbon fixation – turns gaseous carbon – CO_2 – into living organic compounds that grow

carbon sequestration – the removal and storage of carbon from the atmosphere, usually in oceans, forests and soils through photosynthesis

centrifugal forces – forces which drive people, organisations or countries apart

centripetal forces – forces which draw people, organisations or countries together

channel flow – the volume of water flowing within a river channel (also called *discharge, and runoff)

channel storage – water held in rivers and streams

closed system – where there are no *inputs or *outputs of matter from an external source – i.e. where inputs and outputs are balanced

colonialism – where an external nation takes direct control of a territory, often by force

complex river regimes – where larger rivers cross several different relief and climatic zones, and therefore experience the effects of different seasonal climatic events. Human factors can also contribute to their complexity, such as damming rivers for energy or irrigation

convectional rainfall – when the ground warms up, *evaporation takes place and the air above is heated and rises. The rainfall created is often intense and associated with electrical storms and thunder

coral bleaching – when coral turns white because the water is too warm and algae (which lives in the coral's tissues) are ejected

critical threshold – a point beyond which damage becomes irreversible

cryosphere – the frozen part of the Earth's hydrological system

cultural fractionization – measures how diverse countries are, by measuring people's attitudes towards, for example, religion, democracy and the law

culture – the ideas, beliefs, customs and social behaviour of a group or society

D

democracy aid – the allocation of funds to other countries for democracy-building

dependency theory – argues that developing countries remain dependent on wealthier nations, and that their reliance on developed economies is the cause of their poverty

deregulation – the reduction in government involvement in finance and business

Glossary

descriptive statistics – measures such as those of central tendency and dispersion which can describe and be used to show patterns in that data

development aid – financial aid given to developing countries to support their long-term economic, political, social and environmental development

diasporas – dispersed populations away from their homeland

discharge – the volume of water passing a certain point in the channel over a certain amount of time

drainage density – describes whether a river has many or few tributaries. Dense drainage networks have many tributaries and carry water more efficiently

E

economic restructuring – the shift in employment from the primary and secondary sectors into tertiary and quaternary

economic sanctions – financial penalties (such as freezing assets or *trade embargoes) which are designed to put pressure on another country to change their policies or behaviour

ecosystem services – a holistic term to describe the services that ecosystems provide such as soil formation, food provision, climate regulation and recreation facilities

El Niño – a situation occurring every 3-8 years where pressure systems and weather patterns reverse

El Niño Southern Oscillation (ENSO) – the change in air pressure between 'normal' years and *El Niño

enclaves – concentrations of particular communities

energy mix – describes the range and combination of sources required to supply a country with energy

energy pathway – describes the flow of energy between a producer and a consumer, and how it reaches the consumer, e.g. pipeline, transmission lines, ship, rail

energy security – being able to access reliable and affordable sources of energy. These may be domestic, but could also include energy sources from 'friendly' countries

enhanced greenhouse effect – the increase in the *natural greenhouse effect, said to be caused by human activities that increase the quantity of greenhouse gases in the Earth's atmosphere

ethnic – a social group identified by a distinctive *culture, religion, language, or similar

ethnic cleansing – the deliberate removal, by killing or forced migration, of one *ethnic group by another

ethnic segregation – the voluntary or enforced separation of people of different *cultures or nationalities

evaporation – the conversion of water to vapour

evapotranspiration – the combined effect of *evaporation and *transpiration

extraordinary rendition – the secret transfer of a terror suspect, without legal process, to a foreign government for detention and interrogation. The interrogation methods often do not meet international standards, and include the use of torture

F

field (or infiltration) capacity – the maximum capacity of moisture that a soil can hold

flash flooding – when dry soil surfaces become waterlogged very quickly, causing rapid *surface runoff

flux – refers to the movement or transfer of carbon or water between *stores

frontal rainfall – formed when warmer moist air meets colder Polar air. The warmer air is forced to rise over the denser colder air, creating low-pressure and rain

G

genocide – the mass killing of a particular group of people

geo-strategic location theory – a theory developed by Halford John Mackinder which argued that whoever controlled Europe and Asia (the world's biggest landmass) would control the world

geological carbon – carbon which results from the formation of sedimentary carbonate rocks – limestone and chalk – in the oceans

geometric boundaries – borders between countries that have been formed by arcs or straight lines

geopolitical – the influence of geography (both human and physical) on politics, especially international relations

Gini coefficient – used to measure inequality between countries. It uses a figure between 0 (wealth is distributed equally) and 1 (where one person has all the wealth). The higher the coefficient, the more unequal the distribution

Gini index – essentially the same as the *Gini coefficient, except the Gini index is a value between 0 and 100 (i.e. the Gini coefficient multiplied by 100)

gravitational potential energy – ways in which water accelerates under gravity, thus transporting it to rivers and eventually to the sea

groundwater flow – also known as *base flow – slow-moving water that seeps into a river channel through rocks

groundwater storage – water held within permeable rocks (also known as an *aquifer)

H

hard-engineering – human-made, artificial structures designed to protect the land from erosion or flooding

hard power – the expression of a country's will or influence through coercive measures, including *economic sanctions and military force or threat

hegemon – a country or state that is dominant over others

Highly Indebted Poor Countries initiative (HIPC) – a policy introduced by the *IMF and the *World Bank which aimed to reduce national debts of the world's least developed countries by partially writing the debts off, in return for *Structural Adjustment Programmes

Glossary

human capital – the economic, political, cultural and social skills within a country

I

import substitution – boosting domestic manufacturing and production as a substitute for previously imported products

inferential statistics – those which explore relationships between sets of data and can be used to test hypotheses about populations larger than the data set which has been sampled

infiltration – water entering the topsoil. Most common during slow or steady rainfall

input – an input into the system from outside, such as *precipitation into the drainage basin system

Inter-Tropical Convergence Zone (ITCZ) – a narrow zone of low pressure near the Equator where northern and southern air masses converge

interception – temporary storage, as water is captured by plants, buildings and hard surfaces before reaching the soil

intergovernmental organisation (IGO) – an organisation involving several countries working together on issues of common interest

International Monetary Fund (IMF) – a global organisation whose primary role is to maintain international financial stability

J

jet stream – a band of fast-moving air (located 9-16 km above the Earth) which determines the direction of weather systems and their speed of movement

K

Kuznet's curve – the concept that as rapid economic development occurs, environmental degradation increases, but after a certain level of development is reached, action to protect the environment can decrease degradation

L

lag time – the gap between the peak (maximum) rainfall and peak *discharge (highest river level) on a *storm hydrograph

La Niña – when the 'normal' pressure systems and weather patterns intensify and low pressure over the western Pacific becomes lower, and high pressure over the eastern Pacific higher

leaching – the loss of nutrients from the soil by *infiltration

liberalism – the idea that the government's role in business and the economy should be minimal, to allow individual decision-making, a free market and open competition between companies

linear sample – a type of *systematic sampling where a sample is taken at equal points along a line, such as every 10 metres

literature review – the part of the Independent Investigation which involves background reading about your topic and then producing a summary of that information

logarithmic scale – increasing on a cycle, usually by 10, so that 1 to 10 is the first cycle, 10 to 100 is the second cycle, 100 to 1000 the third, and so on

Lorenz curve – used to show and measure inequality in graphical form

M

media plurality – the ownership of several forms of media by the same company

mega-drought – a period of unusually low rainfall, lasting for decades or longer

Millennium Development Goals (MDGs) – goals related to different aspects of human development that were agreed by the UN in 2000, and that were to be achieved by 2015. They included reductions in poverty, hunger and infant mortality

mitigation strategies – strategies which aim to reduce or alleviate the impacts or severity of adverse conditions or events (such as reducing the amount of greenhouse gases that are released)

modernisation theory – a theory that believed that poverty was a trap; traditional family values in poorer countries held economies back; and that capitalism was the solution to poverty

monocultural – consisting of only one *culture

N

national sovereignty – the idea that each nation has a right to govern itself without interference from other nations

nationalise – when a company is transferred from private ownership to ownership or control by the state

nationalism – a patriotic feeling of pride and loyalty to a nation

natural greenhouse effect – the warming of the atmosphere as gases such as CO_2, CH_4 and water vapour absorb heat energy radiated from the Earth

negative feedback – when a change tends to reinforce a system, leading to stability

neo-colonialism – describes how even though less-developed countries may no longer be directly ruled by another, they are still controlled indirectly through economic, cultural and political means

neo-liberalism – a belief in the free flows of people, capital, finance and resources. Under neo-liberalism, State interventions in the economy are minimized, while the obligations of the State to provide for the welfare of its citizens are diminished

non-governmental organisation (NGO) – a not-for-profit organisation, independent from any government

O

ocean acidification – the process of the ocean's pH decreasing as the level of CO_2 in the ocean increases

Glossary

open system – a system with *inputs from and *outputs to other systems

orographic rainfall – when warm, moist air is forced to rise over upland areas, causing the moisture to condense and create rainfall

out-gassing – when volcanoes erupt, releasing terrestrial carbon (held within the mantle) into the atmosphere as carbon dioxide

output – e.g. from a system, such as *evaporation or *transpiration from a drainage basin system

overabstraction – the removal of too much water from groundwater reserves, leading to rivers drying up

P

2015 Paris Agreement – legally binding global climate deal, adopted by 195 countries at the UN's Paris Climate Conference (COP21) in December 2015

percolation – the downward seepage of water through rock under gravity, especially in permeable rocks e.g. sandstone

person sample – a type of *systematic sampling where a person is selected at regular intervals, such as every tenth one on a street

players – individuals, groups and organisations involved in making decisions that affect people and places, known collectively as *stakeholders

point sample – a type of *systematic sampling where a point is selected at regular intervals, such as a series of points located at the intersection of a 10-metre grid

positive feedback – when a change leads to a decrease within a system and creates instability

potential evapotranspiration (PE) – an estimate of the amount of water lost through *evaporation and *transpiration in any given period, depending on temperature and air humidity

precipitation – moisture in any form, such as rain, snow, sleet and hail

primary consumers – organisms such as bugs, beetles, larvae and herbivores which depend and feed on *primary producers, and return carbon to the atmosphere during respiration

primary producers – green plants that use *solar energy to produce biomass

privatisation – the transfer of state or government assets into ownership by private individuals, companies or shareholders

proxy war – a war instigated by a major power that is not always directly involved in the fighting

Purchasing Power Parity (PPP) – (shown in US$) relates average earnings in a country to local prices and what they will buy. It is the spending power within a country, and reflects local costs of living

Q

qualitative – generates information about thoughts and opinions of people

quantitative – generates numerical data

R

radiative forcing effect (RFE) – the amount that a greenhouse gas affects the balance between the Earth's incoming solar radiation and outgoing long-wave radiation

rain-shadow effect – when *orographic rainfall has occurred over an upland area, the area on the lee side of the hills will receive less rain because the air descends, warms and becomes drier

random sampling – when each place or person in an area or a population has a mathematically equal chance of being selected

re-greening – the conversion of dry landscapes to productive farmland

refugees – people who are forced to leave their country because of war, natural disaster or persecution

remittance payments – income sent home by individuals working elsewhere

resilience – the ability of a system to 'bounce back' and survive

rising limb – the line on a *storm hydrograph which shows the *discharge rise up to its peak discharge

river regime – the annual pattern of flow within a river

S

Schengen Agreement – an agreement which abolished many of the internal border controls within the EU and enabled passport-free movement across most EU member states

sectoral sanctions – penalties that are imposed on targeted key areas in another country, such as energy, banking, finance, defence and technology

simple river regimes – where the river experiences a period of seasonally high *discharge, followed by low discharge

smart irrigation – where drip systems allow water to drip slowly to plants' roots through a system of valves and pipes, reducing wastage and *evaporation

soft power – the power that arises from a country's political and economic influence, moral authority and cultural attractiveness

soft-engineering – attempts to work with natural processes in order to mitigate risks

soil moisture – water held within the soil

solar energy – energy from the sun

stakeholders – the collective name for *players who are individuals, groups and organisations involved in making decisions that affect people and places

stem flow – water flowing down plant stems or drainpipes

Glossary

store – an accumulation or quantity of, for example, water or carbon, in a system

storm hydrograph – a graph showing how a river responds to a particular storm. It displays both rainfall and *discharge

stratified sampling – a type of sampling which reflects the population being sampled – e.g. if 35% of the population were aged over 65 then 35% of the people surveyed would be over 65

Structural Adjustment Programmes (SAPs) – policies imposed by the *IMF which forced the State to play a reduced part in the economy and in social welfare, in return for re-arranging loans at adjusted rates of interest, and at more affordable repayments

sub-prime lending – the lending by banks to low-income earners with insecure jobs, who could never normally afford mortgages

superpower – a country with dominating power and influence

surface runoff – flow over the surface during an intense storm, or when the ground is frozen, saturated or impermeable. Also called overland flow

surface storage – any surface water in lakes, ponds and puddles

Sustainable Development Goals (SDGs) – a set of goals, launched in September 2015, to be achieved by 2030 and designed to end poverty, protect the planet and ensure prosperity for all

Sustainable Drainage Systems (SuDS) – techniques such as permeable pavements and infiltration basins which reduce *surface runoff produced from rainfall

systematic sampling – when each sample is selected in a regular manner

T

thermohaline circulation – the flow of warm and cold water that circulates around the world's oceans

throughflow – also known as inter-flow; water seeping laterally through soil below the surface, but above the water table

throughput – the quantity of a material, such as water, that flows through a system or *store

tipping point – when a system changes from one state to another, irretrievably

tonnes of oil equivalent – a unit designed to include all forms of energy by comparing them with oil in terms of heat output. It measures each type of energy by calculating the amount of heat obtained by burning one tonne, and then converting it to however much oil would be required to produce an equivalent amount of energy

trade embargo – a government or international ban that restricts trade with a particular country

transit state – a country or state through which goods or people flow (e.g. energy on its way from producer to consumer)

transpiration – water taken up by plants and transpired from the leaf surface

U

unilateral actions – one country, or group of countries, acting against another, for example, without formal UN approval

V

vegetation storage – any moisture taken up by vegetation and held within plants

virtual water – water transferred by trading in crops and services that require large amounts of water for their production

W

Walker Cell – the circulation of air whereby upper atmospheric air moves eastwards, and surface air moves west across the Pacific, causing trade winds

Wallerstein's world systems theory – a theory which claims that core regions drive the world economy and that peripheral areas (distant from the core and lacking capital) rely on core regions to exploit their raw materials. Therefore, unequal trade develops between them

Washington Consensus – a belief that economic efficiency can only be achieved if regulations are removed

water budget – the difference between *inputs of water (such as *precipitation) and *outputs of water (such as *evapotranspiration) in any given area

water insecurity – the state where present and future supplies of water cannot be guaranteed, caused by *water scarcity and *water stress

Water Poverty Index (WPI) – measures how far a community or country meets all the criteria for: the availability of water resources, access to water, handling capacity, use of water, and the ability to sustain nature and ecosystems

water scarcity – there is less than 1000m³ of water available per person per year. An imbalance between demand and supply of water, classified as physical scarcity (insufficient water to meet demand) or economic scarcity (people can't afford water, even when it's available)

water stress – there is less than 1700m³ of water available per person per year. If a country's water consumption exceeds 10% of its renewable fresh water supply, including difficulties in obtaining new quantities of water, as well as poor water quality restricting usage

World Bank – a global organisation whose role is to finance development

World Trade Organisation (WTO) – a global organisation which looks at the rules for how countries trade with each other

world water gap – refers to the fact that in many parts of the world there is not enough water to meet demand, whereas wealthy countries are consuming greater and greater quantities of water

Revision planner

Date: _____

	Revision period 1	Revision period 2	Revision period 3	Revision period 4	Revision period 5
Monday					
Tuesday					
Wednesday					
Thursday					
Friday					
Saturday					
Sunday					